THE DEFINITIVE GUIDE TO TRANSPORTATION

THE DEFINITIVE GUIDE TO TRANSPORTATION

THE DEFINITIVE GUIDE TO TRANSPORTATION

PRINCIPLES, STRATEGIES, AND DECISIONS FOR THE EFFECTIVE FLOW OF GOODS AND SERVICES

Council of Supply Chain
Management Professionals
and
Thomas J. Goldsby
Dr. Deepak Iyengar
Dr. Shashank Rao

Vice President, Publisher: Tim Moore
Associate Publisher and Director of Marketing: Amy Neidlinger
Executive Editor: Jeanne Glasser Levine
Consulting Editor: Chad Autry
Operations Specialist: Jodi Kemper
Cover Designer: Chuti Prasertsith
Managing Editor: Kristy Hart
Senior Project Editor: Lori Lyons
Copy Editor: Karen Annett
Proofreader: Anne Goebel
Indexer: Erika Millen
Compositor: Nonie Ratcliff
Manufacturing Buyer: Dan Uhrig

© 2014 by Council of Supply Chain Management Professionals
Published by Pearson Education, Inc.
Upper Saddle River, New Jersey 07458

For information about buying this title in bulk quantities, or for special sales opportunities (which may include electronic versions; custom cover designs; and content particular to your business, training goals, marketing focus, or branding interests), please contact our corporate sales department at corpsales@pearsoned.com or (800) 382-3419.

For government sales inquiries, please contact governmentsales@pearsoned.com.

For questions about sales outside the U.S., please contact international@pearsoned.com.

Company and product names mentioned herein are the trademarks or registered trademarks of their respective owners.

Printed in the United States of America

Second Printing June 2015

ISBN-10: 0-13-344909-2
ISBN-13: 978-0-13-344909-9

Pearson Education LTD.
Pearson Education Australia PTY, Limited.
Pearson Education Singapore, Pte. Ltd.
Pearson Education Asia, Ltd.
Pearson Education Canada, Ltd.
Pearson Educación de Mexico, S.A. de C.V.
Pearson Education—Japan
Pearson Education Malaysia, Pte. Ltd.

Library of Congress Control Number: 2013952812

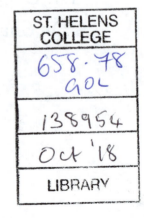

Tom:

To my darling wife Kathie, my two amazing kids Emma and Aiden, my parents Joe and Sujane Goldsby, my brother Mike, my in-laws Doug and Carole Boyd, and great friends, colleagues, and students who motivate and inspire me every day.

Deepak:

To Advay, the joy of my life, my wife Shilpa, parents Mitra and Jawahar, and in-laws Karabi and Chandrashekar, who make life that much more interesting.

Shashank:

To the people who make my life beautiful—my parents Suresh and Shalini Rao, who have strived every day to give me the opportunities they never had. And to the love of my life, Rekha, who understands me better than I understand myself.

CONTENTS

ACKNOWLEDGMENTS

How privileged we are to study and practice logistics and supply chain management. We can think of no other fields in which academia and industry have so much to offer the other. At the forefront of this exchange (for fifty years now) is the Council of Supply Chain Management Professionals (CSCMP). We would like to acknowledge the present and past leadership of the organization for its forthright commitment to advancing our discipline. CSCMP is based on the principles of leading through research and education. To this day, it remains the one place where academics and practitioners simply "must go" to meet, exchange ideas, and collaborate. As academics, we engage in CSCMP activities not only to learn from one another, but to learn from industry and to gain insights into the pressing challenges of modern business. Fortunately, we have something to offer by way of meaningful, real-world research. It makes for a healthy exchange. Sadly, exchanges of this kind between industry and academia are not common to all business fields. We, again, thank CSCMP for fostering such collaboration for the past half century. It is with much enthusiasm that we welcome the next half century.

We also graciously acknowledge Pearson for joining forces with CSCMP in support of the SCPro certification. The certification is an exciting development at the Council, and for the supply chain profession at large. The generation and distribution of world-class materials is only possible with the support of Pearson. Further, we thank the editorial team and Jeanne Glasser Levine, in particular, for serving as excellent stewards of this book series. A special thanks to Dr. Chad Autry for the invitation to offer this work and for keeping us honest in the process.

We would like to acknowledge our home institutions: The Ohio State University, Central Washington University, and Auburn University for their support of our work. These three institutions are dedicated to advancing the field of Logistics and the discipline of Supply Chain Management through the creation of new knowledge, the development of the next generation of business leaders, and our individual efforts to achieve these ends. We are fortunate to have excellent colleagues and collaborators that number too many to list in full, but a short list includes John Caltagirone, Martha Cooper, Keely Croxton, Jim Ginter, Stanley Griffis, Michael Knemeyer, Ike Kwon, Douglas Lambert, Peg Pennington, John Saldanha, Ted Stank, Peter Ward, David Widdifield, and Walter Zinn. A special thanks to Mr. Robert Martichenko of LeanCor for his generous contributions in support of all things that involve learning and good fun. Another huge thanks to the "Columbus Four:" Kenneth Ackerman, Dick Hitchcock, Bud LaLonde, and Art Van Bodegraven for giving so generously to our community and our discipline. We can only hope the others we've undoubtedly omitted here will forgive us and realize that they are, indeed, appreciated.

ABOUT THE AUTHORS

Dr. Thomas J. Goldsby is Professor of Logistics at The Ohio State University. Dr. Goldsby has published more than 50 articles in academic and professional journals and serves as a frequent speaker at academic conferences, executive education seminars, and professional meetings around the world. He is co-author of *Lean Six Sigma Logistics: Strategic Development to Operational Success* and *Global Macrotrends and Their Impact on Supply Chain Management*. He serves as Associate Director of the Center for Operational Excellence, research associate of the Global Supply Chain Forum, and a Research Fellow of the National Center for the Middle Market, all with The Ohio State University. Dr. Goldsby has received recognitions for excellence in teaching and research at Iowa State University, The Ohio State University, and the University of Kentucky. He is co-editor of the *Transportation Journal* and co-executive editor of *Logistics Quarterly* magazine. He serves on the boards for the American Society of Transportation & Logistics and Supply Chain Leaders in Action, the Research Strategies Committee of CSCMP, and as a reviewer for the Gartner Top 25 Supply Chains, LQ Sustainability Awards, SCLA Circle of Excellence Award, CSCMP Supply Chain Innovation Award, and University of Kentucky Corporate Sustainability Award programs. He has conducted workshops and seminars throughout North America, South America, Europe, Asia, and Africa.

Dr. Deepak Iyengar received his Ph.D. in the area of Logistics and Supply Chain Management from the University of Maryland, College Park. His areas of research and teaching include logistics and sustainability in supply chains. He is currently an Assistant Professor at Central Washington University, Ellensburg in the Department of Supply Chain Management. Dr. Iyengar has published works in leading supply chain and logistics journals like *Journal of Business Logistic, International Journal of Physical Distribution &Logistics Management,* and *Transportation Journal* to name a few. In addition, he has mentored students and led projects using the methodology of Lean Six Sigma to various 3PLs.

Dr. Shashank Rao is the Jim W. Thompson Assistant Professor of Supply Chain Management at Auburn University. He holds a B.S. in Environmental Engineering, an M.B.A. in Marketing, and a Ph.D. in Operations and Supply Chain Management. Along with his academic training, Dr. Rao has several years of industry experience in the banking and engineering sectors. He has also served as a consultant to manufacturers and retailers on issues like order fulfillment and distribution management. His research interests focus on retailing, order fulfillment, and logistics customer service. He has published several articles in academic and professional journals of repute such as the *Journal of Operations Management, Journal of Business Logistics, International Journal of Logistics*

Management, International Journal of Physical Distribution and Logistics Management, Industrial Management, Industrial Engineer, and more. He also serves on the Editorial Review Board of the *Journal of Business Logistics,* the *Journal of Supply Chain Management,* and the *Journal of Operations Management,* along with serving as an Associate Editor at the *Journal of Organizational Computing and Electronic Commerce.* He is a frequent speaker at academic conferences, executive education seminars, and professional meetings, and also conducts hands-on training on supply chain technologies like TMS, ERP, and Routing Systems.

Section 1

Transportation: The Basics

1

TRANSPORTATION IN BUSINESS AND THE ECONOMY

Transportation is among the more vital economic activities for a business. By moving goods from locations where they are sourced to locations where they are demanded, transportation provides the essential service of linking a company to its suppliers and customers. It is an essential activity in the logistics function, supporting the economic utilities of place and time. *Place utility* infers that customers have product available where they demand it. *Time utility* suggests that customers have access to product when they demand it. By working in close collaboration with inventory planners, transportation professionals seek to ensure that the business has product available *where* and *when* customers seek it.

Transportation is sometimes to blame for a company's inability to properly serve customers. Late deliveries can be the source of service problems and complaints. Products might also incur damage while in transit, or warehouse workers might load the wrong items at a shipping location. Such *over, short, or damaged* (called OS&D) shipments can frustrate customers, too, leading to dissatisfaction and the decision to buy from a competitor for future purchases.

However, when a company performs on time with complete and undamaged deliveries consistently, this can instill customer confidence and gain business for the company. When a company instills confidence in service performance, it can make customers more reluctant to succumb to competitors' bids to steal business away through clever promotions and reduced prices.

Aside from its service ramifications, transportation can also represent a substantial cost for the business. The cost of transportation can sometimes determine whether a customer transaction results in a profit or a loss for the business, depending on the expense incurred in providing transportation for a customer's order. Faster modes of transportation generally cost more than slower modes. So although shipping an order overseas by airplane is much faster than transporting by ship, it can cost as much as 20 times more. Such a cost difference might not justify the use of the faster way of transporting the goods. Supply chain managers must therefore carefully consider the cost of transporting goods when determining *whether* to move product and *how* to move product in the most economical manner.

This book supports the learning objectives of the Transportation Management module (Learning Block 5) of the Council of Supply Chain Management Professionals (CSCMP) SCPro Level 1 certification. These objectives are stated as follows:

1. Describe the basic concepts of transportation management and its essential role in demand fulfillment.

2. Identify the key elements and processes in managing transportation operations and how they interact.

3. Identify principles and strategies for establishing efficient, effective, and sustainable transportation operations.

4. Explain the critical role of technology in managing transportation operations and product flows.

5. Define the requirements and challenges of planning and moving goods between countries.

6. Discuss how to assess the performance of transportation operations using standard metrics and frameworks.

The book is organized into three sections. Section 1, "Transportation: The Basics," provides a foundation for transportation operations, including a survey of transportation modes, the economics of transportation, and the array of transportation service providers. Section 2, "Transportation for Managers," provides a customer's perspective on transportation, including insights on designing a logistics network, selecting services, and evaluating performance. Content is provided on key aspects of transportation management, including strategy formation, technology deployment, and international supply chain operations. Section 3, "Transportation in 2013 and Beyond," is dedicated to contemporary issues in logistics, including sustainability, and offers an outlook on the future of transportation. Throughout the text, we feature important terms and concepts that are essential for supply chain professionals who are responsible for transportation activity to understand. In this first chapter, we continue by illustrating the role of transportation in the logistics function, the supply chain, and the larger economy.

Transportation and Logistics

Logistics is defined as "that part of supply chain management that plans, implements, and controls the efficient, effective forward and reverse flow and storage of goods, services and related information from the point of origin to the point of consumption in order to meet customers' requirements."[1] Transportation is represented in this expression through the word *flow*. Transportation provides the flow of inventory from points of origin in the supply chain to destinations, or points of use and consumption. Most businesses manage both inbound and outbound logistics. Inbound logistics involves the procurement of materials and goods from supplier locations. Outbound logistics involves the distribution of materials and goods to customer locations. Therefore, transportation is necessary on the inbound and outbound sides of the business.

The definition of *logistics* mentions not only the forward flow and storage of goods, services, and related information, but also the reverse flow.

Inventory sometimes flows in the reverse direction. *Reverse logistics* refers to "the role of logistics in product returns, source reduction, recycling, materials substitution, reuse of materials, waste disposal, and refurbishing, repair, and remanufacturing."[2] So transportation not only delivers material and products to customers, but also moves reusable and recyclable content to companies that can use it. Figure 1-1 shows the forward and reverse flows managed by logistics.

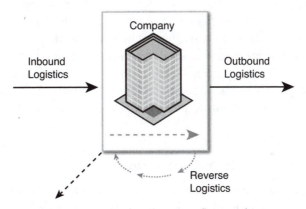

Figure 1-1   Forward and reverse flows in logistics.

Transportation is only one activity responsible for providing time and place utilities through inbound and outbound logistics. Logistics also involves forecasting demand, planning inventory, and storing goods as well as delivering them. Optimized logistics performance means that these activities are working closely together so that the customer of the logistics service is satisfied with the service, yet the cost the company incurs is minimized. This optimal performance requires an understanding of how the various logistical decisions and actions affect service for customers and total cost.

Consider, for instance, that a company seeks to minimize its investment in inventory. The company elects to hold all its inventory in one central warehouse location, for it has been shown that consolidated inventory reduces inventory investment. Warehousing cost should also be minimized because the company is maintaining only a single facility instead of several locations. Customers located close to the central warehouse will be pleased with this decision because the company must travel only a short distance to deliver items to these nearby customers. However, customers located farther from the central warehouse are likely to be disappointed. They will ask for faster transportation to reduce the order lead times. This might involve using faster means of transporting the goods, which, as noted, tends to cost more than using slower modes. In sum, holding inventory in one central location might reduce inventory and warehousing costs, but it will increase transportation costs. The business might also be at risk of losing sales to competitors who can offer shorter and more reliable order lead times.

Conversely, a supply chain strategy that seeks to minimize transportation cost will likely not represent an optimal solution for the company. This might call for shipping orders to customers in large volumes and using slow means of transportation. Requiring large order quantities and using slow forms of transport are two more ways to disappoint customers and risk losing business to competitors. So although transportation is usually a sizeable expense for a company, and often the largest expense in the function of logistics, supply chain managers must consider the interrelationships among the various logistical actions and costs. Tradeoffs are often associated with these decisions, and the company's customers are also affected. The recognition of interrelationships among transportation, inventory, warehousing, information exchange, and customer service is the embodiment of a *systems approach*.[3] The manager seeks to optimize the performance of the logistics system instead of optimizing a singular element of the system. This book, therefore, treats transportation as one important element of the logistics system.

Transportation and the Supply Chain

Another system that calls for recognizing tradeoffs and interrelationships among actions and costs is the supply chain. A *supply chain* is the network of companies that work together to provide a good or service for end users and consumers. Most companies operate within supply chains, relying on outside parties such as suppliers and customers to help them reach the end-user market.[4] In other words, most companies do not entirely own their supply chains.[5]

Supply chain management encompasses the planning and management of all activities involved in sourcing, procurement, conversion, and logistics management. It also includes coordination and collaboration with channel partners, which can be suppliers, intermediaries, third-party service providers, or customers. Supply chain management integrates supply-and-demand management within and across companies.[6] Managing a supply chain, then, means managing the business relationships among the focal company and its outside supply chain partners, including customers and suppliers.[7]

As Figure 1-2 illustrates, transportation represents the physical connection among the companies in the supply chain. The locations in a supply chain network are called *nodes*, and the connections are referred to as *links*. When one entity sells product to another, transportation provides the delivery. An outbound delivery for a supplying company is the inbound delivery for its customer. When one level in the supply chain experiences delays and problems, it impacts the abilities of downstream members of the supply chain to serve their customers. For this reason, the larger economy is affected when transportation disruptions occur. Potential sources for disruptions include equipment failures, natural disasters and inclement weather, work stoppages, and government intervention. The next section reviews the role of transportation in the larger economy.

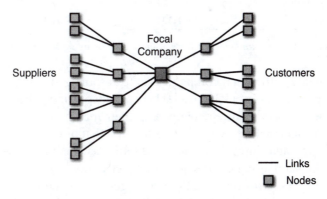

Figure 1-2   Links and nodes in a supply chain.

Transportation and the Economy

The business of moving freight is a major expense for an individual company and is essential for flowing product through the supply chain. In total, transportation is a significant industry in every developed economy. Each year, the CSCMP conducts an analysis of logistics costs in the United States. Table 1-1 illustrates the expenditures directed toward various logistics activities in the United States in 2012. Of the $1.331 trillion sum, $836 billion (or 62.8 percent of total logistics cost) was spent on transportation. This amount greatly exceeds the expenses dedicated to other logistics activities. Logistics costs amount to 8.5 percent of the nation's gross domestic product (GDP), and transportation alone represents 5.4 percent of the U.S. GDP. In other words, just over 5 cents of every dollar spent in the United States goes toward transportation. Total logistics costs in Europe run the range of 12 percent. In less developed countries, the share of the GDP directed toward transportation can be even greater because it costs more to move products when infrastructure is lacking or not sufficiently maintained. Total logistics costs in China are approximately 21 percent of the GDP.

Table 1-1   U.S. Logistics Cost, 2012

Cost Category	$ Billions	% of Total
Inventory-Carrying Cost	305	22.9
Transportation	836	62.8
Warehousing	130	9.8
Shipper-Related Costs	10	0.8
Logistics Administration	51	3.8
Total Logistics Cost	**1,331**	**100**

Source: Rosalyn Wilson (2013), *The 24th Annual State of Logistics Report: Is This the New Normal?* CSCMP.

The ease or difficulty with which companies can transport goods within a country can affect their competitiveness in global trade. When transporting goods is easy and costs are relatively low, exporters can efficiently ship their merchandise to export locations and on to international markets. One such example is that of U.S. farmers in the central states of Illinois, Indiana, Iowa, Nebraska, and Ohio. Farmers in these states compete with farmers in the Pampas region of Argentina to sell grains, such as corn and soybeans, in markets abroad. The growing conditions in Argentina are considered advantageous to those in the United States, allowing farmers to achieve greater yields and enjoy lower costs of production. However, by virtue of using the highly efficient U.S. railroads and river barges to reach the export port, the American farmer enjoys savings in transit time and transportation expense that can offset the inherent advantages in production yield and costs enjoyed by the Argentinian farmer. The more difficult it is to move product over a distance, the greater the *friction of distance*. With greater friction come higher costs. In the grain-shipping example, the farmer in Argentina faces greater friction of distance (and higher costs) in transporting the grains from the farm to the export port in Buenos Aires. Despite the longer distance, there is less friction for the American farmer, who can efficiently ship the grains via railroad and barge.

Table 1-2 contains statistics of the transportation infrastructure in the United States. With more than 4 million miles of public roadways, enough to circle the globe 157 times, virtually every business and household in the nation enjoys the benefits of easy roadway access. The Interstate Highway System (originally called the National System of Interstate and Defense Highways) provides efficient connectivity among almost every large and medium-size city in the nation. High-speed delivery is supported with a network of more than 5,000 public airports, as well as another 14,339 for private use and 271 for military purposes. Freight rail transportation occurs over 161,000 miles, most of which is operated by major Class I rail operators. The continental United States is also endowed with various forms of navigable waterways, including rivers, Great Lakes, and ocean shipping on three coasts. Finally, pipelines, an often overlooked mode of transportation, distribute large quantities of fluid material (gas and liquids) over long distances throughout the nation. In sum, this extensive network of transportation assets supports commerce among businesses and consumers within the United States, and also helps to support export and import activity with businesses abroad.

Table 1-2   U.S. Transportation Infrastructure

Miles of Public Roadways	4,059,339
Miles of Interstate Highway System	47,011
Miles of National Highway System	117,084
Miles, Other Roads	3,895,244
Number of Public Airports	5,172
Miles of Railroad	161,195
Miles of Navigable Waterways	25,320
Number of Commercial Ocean Facilities	5,588
Miles of Pipelines	1,735,237
Miles, Hazardous Liquid Pipelines	177,631
Miles, Gas Transmission and Gathering	324,606
Miles, Gas Distribution	1,233,000

Source: These data are presented in the *2013 Pocket Guide to Transportation*, Bureau of Transportation Statistics, U.S. Department of Transportation, compiled from various sources.

A strong argument can be made that the economic health of a nation is linked to the health of its transportation infrastructure. This argument works in two ways. First, an extensive infrastructure supports economic growth. Second, economic strength supports investment in infrastructure. A modern case study that illustrates the strong association between infrastructure development and economic growth is that of China during the past two decades.

Transportation supports an economy not only by connecting people and places, but also through the many people it employs. Table 1-3 presents employment data for the United States in 2011. Nearly 4.3 million people were employed in the provision of for-hire transportation services, with almost 1.3 million in the trucking industry alone. Another 5.4 million were employed in transportation-related services and construction. Finally, almost 1.7 million people were employed in the production of transportation equipment in the United States. Combined, this totals 11.4 million transportation-related private sector jobs, or almost 9 percent of the total U.S. labor force.

Table 1-3   U.S. Transportation-Related Employment, 2011

For-hire transport and warehousing	4,292
Air	456
Water	63
Railroad	229
Transit/Ground Passenger Transportation	436
Pipeline	43
Trucking	1,299
Support Activities	564
Scenic/Sightseeing Transportation	29
Couriers/Messengers	529
Warehousing/Storage	646
Related Services and Construction	5,405
Transportation-Related Manufacturing	1,684

Source: *2013 Pocket Guide to Transportation*, Research and Innovative Technology Administration, Bureau of Transportation Statistics, U.S. Department of Transportation, Washington, D.C.

Transportation and logistics are regarded as "derived market" activities. That is, demand for transportation and logistics service is derived from the demand of other goods and services in the economy. When a manufacturer seeks supplies from distant locations, there is demand for transportation. Similarly, when consumers have demand for goods produced elsewhere, transportation is demanded. So demand for transportation tends to closely follow the economic activity in a region. For this reason, economists and market analysts pay close attention to transportation shipment data—they present an accurate, timely picture of economic vitality for a region or nation. Rail and trucking volume reflect the economic activity of a nation, and ocean and air transportation statistics illustrate trade levels among nations.

Transportation, Society, and the Environment

Thus far, we have explored the many ways in which transportation contributes to the economic health of individuals, businesses, and entire nations, by facilitating the flow of commerce and providing employment opportunities. Transportation can impact our lives in other profound ways, however. In some ways, it can be lifesaving; in others, it is damaging. In this section, we explore some of the different ways transportation affects society and our physical environment.

In the case of emergency situations and calls for humanitarian relief, transportation is essential to supporting lifesaving missions. Such is the case following a natural disaster.

When earthquakes, hurricanes, floods, or other events imperil a region, the ability to deliver supplies of water, medical equipment, communication equipment, and energy is most pressing. In recent years, relief organizations have adopted advanced transportation methods to support the ability to deploy crucial resources to those in need whenever and wherever a crisis arises.

Transportation also plays a critical role in the success of military endeavors, dating back to the days of Thutmose III and his storied conquests in the fifteenth century B.C. that transformed Egypt into a "superpower." The same assertion holds true today. The timely deployment of soldiers, armaments, and supplies has often been credited with the success or failure of military campaigns and peacekeeping missions. A considerable share of the budget for any substantial military body is the provision of transporting soldiers and supplies.

Transportation activity also results in many unintended consequences on society. It is one activity that interfaces directly with people in their day-to-day lives. Manufacturing, another major economic activity, occurs within the confines of buildings. Transportation, on the other hand, involves roadways that passenger vehicles share. More than 32,000 fatalities and 2.2 million injuries occurred on U.S. highways in 2011, with about 1 in 10 fatalities attributed to collisions involving large trucks.[8] Fortunately, these numbers represent steady declines over the past several decades. The declines can be attributed to immense safety improvements in passenger vehicles, increased use of safety belts, and fewer incidents involving alcohol. Sharing the roads with 80,000-pound tractor-trailers and crossing railroad grades remain hazardous though.

Transportation is also the biggest consumer of energy resources in our economy, easily outpacing manufacturing and consumer household usage. According to the Department of Energy, the U.S. transportation system consumes more than 13 million barrels of petroleum each day. This also leads to transportation's regrettable role as the largest contributor to greenhouse gases. Transportation is responsible for emitting nearly 2 billion metric tons of carbon dioxide in the United States each year. Growing concerns of sustainability are changing the way many companies elect to ship products and also the operations of the transportation providers themselves. The carbon footprint for a shipment, an estimate of the greenhouse gases emitted, has become a critical measure of transportation performance, alongside transit time and cost. Consumers are demanding sustainable products and cleaner ways of transporting them. Since 2009, UPS has offered a carbon-neutral service: The shipping company offsets the carbon emissions associated with package delivery by investing in carbon-reduction projects around the world.

Finally, in light of its central role in economic activity, transportation is also a common target for terrorist activity. Whether one considers the hijacking of passenger airlines on September 11, 2001; the disruption of pipelines in oil-producing countries; the use of vehicle-borne improvised explosive devices (VBIEDs) or "car bombs"; or the hijacking of ships off the coast of Africa, transportation assets can be particularly difficult to secure and, thereby, vulnerable to attack. The formation of the Transportation Security Administration (TSA) immediately following the terrorist attacks of September 2001 serves as testament to the intense focus placed on security and safety in transportation activity. To date, much focus has been directed to passenger safety, although freight transportation has seen greater regulation and scrutiny of shipments. This trend is expected to continue into the future.

Summary

These final observations illustrate the many and diverse ways that transportation impacts our lives. Transportation is a major contributor to the economy and a competitive force in business. It is the activity that physically connects the business to its supply chain partners, such as suppliers and customers. Furthermore, the service rendered through transportation activity is a major influence on the customer's satisfaction with the company.

Key takeaways from this chapter include:

- Transportation helps to fulfill the economic utilities of place and time.

- The cost and service aspects of transportation decisions must be balanced with the inputs of inventory, warehousing, order processing, information, and customer service policies to serve customers at the lowest possible cost.

- Transportation provides the links that connect nodes in the supply chain network, linking a focal company to its suppliers and customers.

- Transportation is a major contributor to economic prosperity for a nation. The more efficient the transportation system, the easier it is to conduct commerce.

- Although transportation is an essential activity in any economy, it presents several hazards to society, including risks of accidental injury and death, greenhouse gas emissions, environmental impact, and terrorist activity.

Endnotes

1. Definition offered by the CSCMP.

2. James R. Stock (1998), *Development and Implementation of Reverse Logistics Programs*, Council of Logistics Management (CLM), Oak Brook, IL.

3. For a complete treatment of the total cost of logistics and the tradeoffs among logistics activities and costs, see: Douglas M. Lambert, James F. Robeson, and

James R. Stock (1978), "An Appraisal of the Integrated Physical Distribution Management Concept," *International Journal of Physical Distribution & Logistics Management* 9 (no. 1): 74–88.

4. An end user might be a consumer, business, government entity, or nonprofit organization. It represents the person or organization at the end of the supply chain who will put the supply chain's product or service to use.

5. Exceptions to this observation do exist. Very large oil and gas companies, for instance, often own the sources of supply, the processing of crude into refined oil and gasoline, and the distribution channels of retail refueling stations.

6. Definition offered by the CSCMP.

7. Martha C. Cooper, Douglas M. Lambert, and Janus D. Pagh (1997), "Supply Chain Management: More Than a New Name for Logistics," *International Journal of Logistics Management* 8 (no. 1): 1–14.

8. *2013 Pocket Guide to Transportation*, Research and Innovative Technology Administration, Bureau of Transportation Statistics, U.S. Department of Transportation, Washington, D.C.

2

A SURVEY OF
TRANSPORTATION MODES

Figuring out how to transport goods is among the most fundamental business decisions a company can make. In some cases, the means of moving product is limited as a function of availability or the geography over which a shipment must move. Shipments by truck and rail are primarily limited to land, for instance. Navigable waterways must be available for ships, and takeoff/landing strips must be available for air transportation to be viable. Where multiple options are available, the logistics manager must evaluate the service attributes and costs associated with available options.

Chapter 1, "Transportation in Business and the Economy," introduced the five modes of transportation: road, rail, water, air, and pipeline. This chapter reviews each of the modes and compares their relative service and cost characteristics. Attention then shifts to how the benefits of different modes can be combined through intermodal transportation.

An Overview of the Modes

Chapter 1 noted that $836 billion was spent on transportation in the United States in 2012. Figure 2-1 illustrates the modal split for this sum. The term *modal split* refers to the market share, as measured by total revenues earned by carriers operating within that mode. The dominant share illustrated by trucking (road) transportation is quite clear, with 77 percent of revenues directed toward this one mode. The ease with which shipments can be collected and delivered by road explains much of the success of truck transportation. Yet this mode is also relatively fast, dependable, and flexible in terms of the types of cargo and volume that it can carry. Rail transportation is a distant second, at 9 percent. Note that the rail system in the United States is well developed and competitive, with more than 161,000 miles of track. Similarly, the modes of air, water, pipeline, and forwarders[1] have extensive capacity and offer comprehensive services in the United States. Quite simply, most goods move by truck at some point in their distribution.

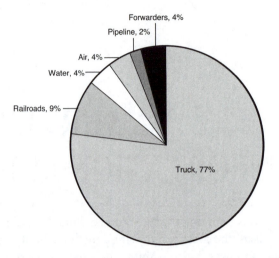

Forwarders, 4%
Pipeline, 2%
Air, 4%
Water, 4%
Railroads, 9%
Truck, 77%

Source: Rosalyn Wilson (2013), *The 24th Annual State of Logistics Report*: Is This the New Normal? Council of Supply Chain Management Professionals.

Figure 2-1   U.S. domestic modal split (revenues).

When viewing modal split under a different measure, however, we see a different story. A conventional unit of measure used in transportation is the *ton-mile*, a combined measure associated with moving 1 ton (2,000 pounds) a distance of 1 mile. Figure 2-2 reveals that although the United States relies greatly on truck transportation for moving freight domestically, the nation also makes extensive use of several other modes. Notably, rail transportation just surpasses trucks when comparing ton-miles. How is this possible? Rail transportation is recognized as extremely efficient for moving large volumes of freight over long distances. The combination of large volumes and long distances allows rail to exceed the high-frequency movements completed by truck, which, on average, are of much lower volume and shorter distances. Pipeline is another mode that excels in moving large volumes over long distances. Although pipelines represent only 2 percent of the transportation market by revenue (in Figure 2-1), they accommodate 15 percent of the ton-miles. Multimodal transportation, referring to a combination of different modes for supporting transportation needs, represents 11 percent of the ton-mile share. Water finds balance in representing 4 percent of revenues and 4 percent in ton-miles. Air transportation occupies a very small share of the domestic modal split in ton-miles. As a premium mode of transportation capable of covering long distance very quickly, it excels in international transport.

When compared side by side, the transportation modes indicate distinct differences in key aspects of service and cost. The balance of this chapter reviews each mode in detail.

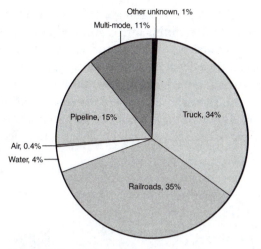

Source: U.S. Department of Transportation, Bureau of Transportation Statistics, *Commodity Flow Survey* data (2009)

Figure 2-2   U.S. domestic modal split (ton-miles).

The Five Modes of Transportation

Transportation occurs in five different ways, based on the commercial situation at hand. These are reviewed in the following sections.

Road Transportation

Road (or motor) transportation is the most common form of transport in most settings. Given the extensive network of roadways within towns and cities, as well as the connections among them, most origin–destination pairs within a land mass can be reached via some means of motor transportation. Motor transportation includes a wide range of roadway transport, such as trucks, vans, cars, or motorcycles. Trucks serve as the primary means of moving shipment of a single pallet-size container to several pallets. A 53-foot trailer, which has become quite common in many settings, can hold two rows of 13 pallets, or 26 pallets. If the goods can be double-stacked (one pallet atop another), the capacity doubles to 52 pallets. The weight capacity for such a trailer is in the range of 45,000 pounds.[2] For smaller loads or city deliveries, smaller vehicles such as box trucks and vans are preferred, as seen in Figure 2-3.

Transportation by road is also a relatively fast and reliable means of transporting goods. Although road travel is not as fast as air transportation, trucks are faster than the other modes. When moving freight from one city to another, trucks can usually average approximately 50 miles per hour, including stops. In a 10-hour shift, this means that a truck can cover approximately 500 miles in a single day. Some companies are employing strategies that can dramatically stretch this range. Among these strategies is the use of "team drivers," or assigning two drivers to a truck and allowing them to alternate driving duties. This arrangement can virtually double the daily range of a truck. Another strategy is to devise relay networks that allow a driver to cover the typical 500 miles in a shift and then to hand the truck or trailer to another driver, who then continues with the delivery toward its destination. Although these strategies help to improve range, increasing congestion on roadways, particularly in cities, is challenging the speed and reliability of transport by truck.

Figure 2-3   Box truck for city delivery.

Aside from convenience and speed, road transportation is common for other reasons. The availability of motor carriers contributes to its popularity. For instance, the United States alone has more than 725,000 registered motor carriers. Trucking companies compete vigorously for business in such a competitive market.[3] The competitiveness of the trucking market has accelerated as a result of deregulatory actions and market freedoms instituted over the past 35 years.

The U.S. interstate trucking industry was largely deregulated on matters of economic competition in 1980 through provisions of the Motor Carrier Regulatory Reform and Modernization Act (known as Motor Carrier Act, or MCA-80). Until passage of this historic law, the Interstate Commerce Commission (ICC) regulated matters of market entry and pricing. From the time of its formation in the Act to Regulate Commerce in 1887, the

ICC enjoyed powers that were quasi-legislative, executive, and judicial in orientation. In other words, the ICC was regarded as the single most influential body for all economic and competitive matters associated with surface transportation in the United States, including the trucking industry. MCA-80 removed most of its powers, and carriers began to operate in a free market-based system for matters of interstate transportation. Note that intrastate trucking (transporting from an origin to a destination within the same state) remained regulated by individual states and their motor rate bureaus until 1994, when states were stripped of this power.[4]

Although the MCA-80 legislation applied only to trucking operations within the United States, other nations have followed with policies that liberalize competition in the trucking and transportation marketplaces. Many carriers filed for bankruptcy and closed as a result of the intensifying competition resulting from MCA-80 in the United States, but several other new businesses entered the trucking market because of greatly reduced barriers to entry. Furthermore, MCA-80 is credited with instilling an atmosphere that fostered creativity and innovation in service offerings that was lacking in a staunchly regulated environment. Deregulation allowed for greater differentiation of service, along with competitive pricing, which greatly benefited shippers. Closer collaboration between shippers and their carriers fostered strategies that reduced logistics costs and allowed shippers to better leverage carrier capabilities. In fact, many shippers discontinued the use of private fleets (self-owned and operated trucking fleets) when they determined that they could buy enhanced services at competitive prices in the open marketplace. Shippers in other nations observed the benefits enjoyed by U.S. companies and pushed effectively for greater liberalization in transportation markets in their nations. This trend continues today.

Trucking Market Segments

The trucking market is divided into three broad segments. These include segments based on volume (less-than-truckload and truckload), geographic coverage (regional, national, and international), and equipment (dry van, refrigerated, flatbed, tank, and special equipment).

Truckload carriers are the largest segment in the road transportation market. As their name implies, truckload carriers specialize in moving large volumes of freight for their industrial customers. Although truckload carriers do accept smaller loads, they typically target shipments in the range of 15,000 to 50,000 pounds. Truckload carriers specialize in door-to-door service, collecting freight at an origin and delivering directly to the destination without any intermediate stops. This "one-touch" service is attractive to shippers because it reduces the time in transit and the propensity for damaging shipments by avoiding the rehandling of freight. The larger the geographical area that the truckload operator seeks to serve, the more likely the carrier will have multiple terminal locations for housing equipment and station drivers. Many large nationwide carriers also operate maintenance facilities to service and maintain their equipment at several terminal locations. Aside from these concentrated expenditures on facilities, the capital expenditures for truckload businesses tend to focus on the trucks (also referred to as tractors or power units) and trailers. Each truck should have a trailer, but it is not uncommon for a large truckload business to keep three or four trailers for each truck in their fleet, to ensure that a sufficient number of trailers is available to serve customers. This is particularly true when carriers offer "drop trailer" service, which means that the carrier allows a shipping customer to keep a trailer for an extended time for loading or unloading purposes.

Larger carriers tend to have more capital dedicated to terminal and maintenance facilities, as well as a higher trailer-to-truck ratio. However, it is possible for small carriers to survive with a single truck and trailer. A new Class 8 (heavy-duty truck with gross vehicle weight of 33,000 pounds or more) might cost up to $200,000. But used equipment can cost significantly less, allowing smaller players to enter the market. For this reason, most of the 700,000-plus registered motor carriers on record are small operators, with one or a few trucks composing a fleet.[5] For instance, farmers in North America commonly operate small truckload businesses during the winter (off-season) months for the agriculture business. In season, these carriers use their fleet to provide transportation support for the farm.

The key to survival for any truckload business, large or small, is to make the best possible use of the available equipment and the driver's time. These companies seek to have loaded "revenue miles" to the greatest extent possible. When a truck is not loaded with freight, it still incurs the costs of operation (such as fuel, asset depreciation, and sometimes driver wages) on what are called "deadhead miles." Therefore, it is essential that these carriers find revenue opportunities not only for fronthaul movements (origin to destination moves), but also on backhauls (from the destination back to the origin). The high level of competition in the truckload market requires that prices include only small margins for carriers. One-way hauls with deadhead backhauls are usually not sufficient to support a truckload business.

Less-than-truckload (LTL) carriers specialize in smaller loads, typically in the range of 150 to 20,000 pounds. Whereas truckload carriers typically dedicate a trailer to a customer for an origin–destination movement, LTL carriers operate under the premise of sharing the trailer's capacity among multiple shippers. As a result of this "shared capacity" premise, LTL carriers tend to employ a very different business model than their truckload counterparts. As noted, truckload carriers operate terminals that serve the purpose of stationing drivers and equipment. LTL carriers, on the other hand, employ facilities for collecting and sorting freight, as well as domiciling equipment and drivers.

The difference in operations between truckload and LTL carriers influences costing and pricing for these classes of service. The truckload carrier collects a shipment at a shipper (*consignor*) location and delivers the load directly to the customer (*consignee*) location. The LTL carrier collects freight not only at the shipper location, but also from multiple shippers located throughout the city on the driver's assigned route. After collecting freight at multiple shipper locations, the truck returns to the pickup terminal, where it joins with other trucks that collected freight throughout the day. The trucks are then unloaded, and the freight is sorted. Deliveries destined for a common city or region are loaded onto a *linehaul* truck for transit to a delivery terminal. It is sometimes necessary to direct freight through another intermediate location, called a *breakbulk facility*, for further consolidation.

When freight arrives at the delivery terminal, it is unloaded again, sorted, and loaded onto delivery trucks designated for different routes throughout the city or region. Along one of these routes, a driver completes the delivery to the customer location. LTL drivers often not only deliver freight on these routes, but also collect freight along their routes. In total, LTL deliveries usually require at least one overnight time period to perform the pickup, collection sort, linehaul, delivery sort, and ultimate delivery sequence. For this reason, LTL deliveries usually require more time to complete than truckload deliveries.

Also worth noting is that providing LTL service on a large scale requires considerable capital investment. Terminals and breakbulk facilities add to the expense of operating in this market segment. For this reason, the LTL market tends to be much more concentrated than the truckload market, with fewer large LTL carriers enjoying sizeable shares of the market.

The second market segment considered here is the geographical scope of coverage. Some carriers choose to operate in a smaller market on a regional basis. Others provide coverage for an entire nation. Still others provide trucking services that cross national boundaries. In the days of a regulated trucking industry in the U.S., carriers commonly specialized in service within a single state. Today, very few carriers operate on such a limited basis; instead, they seek to provide broader market coverage. Even regional carriers usually maintain relationships with other regional carriers so that they can market and sell broader coverage to shipping customers, allowing the carriers to compete with nationwide service providers. These inter-regional operations are usually managed on a reciprocal basis among the regional carriers, with parties seeking to achieve balance in the freight that each party is contributing to the combined business.

Similar arrangements are often made on international freight. In these situations, a carrier specializing in one service in one nation hands off cargo to a partner firm in the destination nation. Individual carriers that provide service in more than one nation must be licensed to operate in each of the various settings. *Cabotage laws*, however, are protective regulatory measures that limit the access that non-native providers can offer in a foreign nation. The North American Free Trade Agreement (NAFTA) has provisions designed to allow trucking companies to more freely operate across the national boundaries of Canada, the United States, and Mexico, although these provisions are not yet fully observed. The European Union (EU) allows much greater flow of trucks throughout its member nations.

The third market segmentation of interest is that of equipment. The typical trailer used in truckload and LTL services is the dry van trailer. This trailer uses a conventional enclosed "box" design and has no form of refrigeration or climate control. Temperature-controlled carriers, on the other hand, employ insulated trailers equipped with heaters or refrigeration equipment. Food products often require refrigeration or freezing in transit to maintain the integrity of products. Chemical products and advanced technologies sometimes require refrigeration to avoid being exposed to high levels of heat. Most temperature-controlled carriers also provide dry service. However, few dry carriers are in the temperature-controlled business.

Another common form of equipment is the flatbed trailer. Open-air flatbed equipment is often used to haul industrial equipment and building materials (such as timber, lumber, steel, and pipe) that can be secured with straps and that require little or no protection from the natural environment. Hopper trucks are often open-top trucks and trailers that are used to haul grains (such as corn, soybeans, and wheat), construction materials (such as sand and gravel), and coal. Tank trucks are used to ship fluid materials such as water, oil, and corn syrup. Carriers might also specialize in the areas of household goods transportation, autohauling, cement, or oversize loads. Figures 2-4 through 2-7 show several of these different forms of equipment.

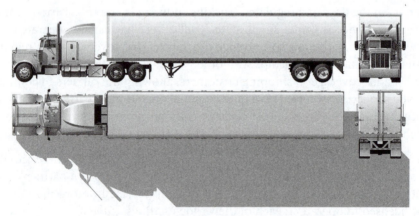

Figure 2-4   Truck and dry van trailer.

Figure 2-5   Truck and tank trailer.

Figure 2-6   Truck and flatbed trailer.

Figure 2-7   Truck with autohauler trailer.

Discussion to this point has focused on industrial (business-to-business) forms of roadway transportation. Consumers, households, and small businesses are also served via roadway transportation. Instead of using large trucks, however, deliveries for these purposes involve box trucks (known as straight trucks or city trucks) for furniture and household appliances, conventional vans for flowers and rugs, and cars for books and pizzas. Motorcycles and bicycles are used for deliveries in urban areas because they can quickly navigate crowded streets. Off-road motorcycles are also used to deliver critical supplies in remote locations that lack good roads.

The diversity of the forms of carriage in roadway transportation speaks to the dependence of both businesses and consumers on this mode. Among industrial shippers, trucking represents about 85 percent of the transportation industry. Road transportation is also the primary means small businesses and consumers employ. With the tremendous growth in Internet retailing, success in home delivery is expected to become critical to businesses in the future. More discussion of this topic is found in Chapter 10, "The Future of Transportation."

Rail Transportation

Another form of "surface" transportation is railroads. Railroads offer the advantage of efficiently transporting very large volumes over long distances. A typical tractor-trailer truck might have a capacity of up to 50,000 pounds of cargo, but a single rail boxcar can carry approximately three times this volume. Furthermore, boxcars are usually grouped in large numbers to form a train, often with more than 100 cars. This speaks to the efficiency with which trains can transport goods.

The infrastructure for rail transportation is not as extensive as the road network, making access something of a challenge. The United States has more than 4 million miles of roadways, but the total rail network totals just over 161,000 miles. Furthermore, this number has declined dramatically since the railroad industry saw dramatic deregulation in 1980 (the same year as MCA-80). Similar to interstate trucking, the freight railroad industry was heavily regulated by the ICC for nearly a century. The Staggers Rail Act, enacted just one month after the MCA-80, more freely allowed rail carriers to determine their destinies by allowing greater pricing scrutiny and reducing scrutiny for railroad abandonment. That is, if a rail operator found a rail segment to be unprofitable, the carrier could elect to close that segment. Until the Staggers provision, rail carriers could petition the ICC for the rights to abandon rail, but the ICC could force the carrier to continue the service under the provision of public interest. This power was greatly reduced under Staggers, and railroad operators abandoned many miles of their networks that they deemed as underperforming financially. Smaller local or regional rail operators sometimes acquired the abandoned lines. Today, many unclaimed lines have been converted into recreational bike paths under the Rails to Trails Conservancy.

The role of railroad transportation has been reduced from 150 years ago, when it served as an instrument of westward expansion in the United States. The Golden Spike ceremony of May 10, 1869, when the Central Railroad and the Union Pacific Railroad met in Promontory, Utah Territory, marked a major milestone in American history as the West joined the East with the first transcontinental railroad. The railroad industry flourished in the back half of the 1800s and early 1900s. In fact, the number of track miles peaked in the United States in 1916, with 254,251 miles. The railroads were prominent in transporting people as well as freight among the growing population. In fact, many small cities were formed along the routes and intersections of the railroads. The advent and growth of automobiles greatly reduced dependence on railroads. With expansions in the roadway network and advances in automobile design and performance, passenger and freight traffic diverted from the railroads to roadways throughout much of the twentieth century.

The ability to provide door-to-door service for virtually every origin–destination within a land mass is one advantage of roadway transport over railroads. The average speed and reliability of delivery are two performance criteria that tend to favor roadways over rail. As noted, railroads seek to combine many railcars to form trains that can move long distances extremely efficiently. It can take several days for a train to form among railcars provided from many different shippers. An individual railcar also might travel with multiple trains, stopping in *switchyards* to transfer from one train to another before it reaches its destination. Each transfer can add days to the transit. For these reasons, transportation by rail can take considerably longer than by truck, with greater variability in the time required to complete the transit. An exception to these observations occurs when a company hires a rail carrier for dedicated service between an origin and destination. Large shippers, such as coal companies, grain marketers, and automakers, often form dedicated trains, or *unit trains*. The average speed of transit is greatly improved when bypassing switchyards and moving direct to destinations.

The countervailing force to the longer transit time of trains is the immensely efficient means of moving freight. As noted, railroads thrive on the ton-miles measure because they are designed to efficiently move large volumes over long distances. One large U.S. railroad estimates that it consumed roughly 490 million gallons of fuel to move just over 229 billion ton-miles of freight in 2010. This equates to approximately 467 ton-miles per gallon of fuel. By comparison, the carrier claims that a truck would require 71 gallons of fuel to move 19 tons over a distance of 500 miles, equating to about 134 ton-miles per gallon.[6] On these grounds, the railroad claims to be 3.5 times more fuel-efficient than trucking operations. On a related note, the Association of American Railroads estimates that trains emit 75 percent fewer greenhouse gases than trucks, when moving comparable volumes and distances. These are important observations in the age of energy conservation and sustainability.

Rail Transportation around the World

Rail transportation is not as prominent for moving freight in settings outside the United States. Several reasons are offered for this distinction. During the second half of the nineteenth century, when rail was instrumental in extending development and establishing trade outposts and new cities in the western United States, many other nations were more developed, with established trade routes. The United States, on the other hand, was leveraging the capability of emerging steam-power locomotives to forge growth and development.

In addition, instead of focusing on freight movement, several nations employ rail for passenger purposes and often heavily subsidize the construction and operation of these services. Passenger trains are typically called on for only limited freight service, often for the transport of mail and parcels. This is true in several European and Asian nations, where high-speed rail garners much attention for its marvels in moving people at speeds of up to 185 miles per hour (300 kilometers per hour). China and South Korea are currently the only operators of magnetic levitation (or "maglev") technology for commercial rail purposes, although many other nations are developing these technologies that allow trains to travel on magnets instead of using conventional rails and wheels.

Finally, rail is a mode that excels for movements of greater than 500 miles.

Such long-distance movements would exceed the national boundaries of many nations, requiring coordination across nations to leverage its potential. Yet many nations remain cautious about establishing connections across national boundaries via rail links. This was particularly true in the boom times of railroad development, before widespread automobile usage. Nations tended to be much more insular from an international trade perspective than today. In fact, where rail infrastructure developed, nations built rail lines of differing gauges (track widths), requiring people and goods to physically transfer when transported from one nation to another.

Two developed nations with large land masses, China and Russia, join the United States as major freight rail nations (see Table 2-1). With considerably less usage than the top three, India, Canada, and Brazil follow. This is not to suggest that freight rail is an insignificant mode in any setting in which it operates. For instance, Germany uses rail extensively to transport cargo from its northern port in Hamburg to destinations in the south of the country. Private investment also is starting to develop rail transport in several nations of South America and Central Europe, which will spur considerable usage.

Table 2-1   Top Ten Nations for Freight Rail Volumes (in Ton-Miles, Billions)

China	1,830
United States	1,534
Russia	1,249
India	415
Canada	201
Brazil	166
Ukraine	135
Kazakhstan	122
South Africa	70
Germany	66

Source: International Railway Union, Railisa, the UIC Statistics Database.

The cost of building rail lines can make them prohibitively expensive in many settings. In addition to the rail line itself, the operator must procure the land upon which the rail is to be placed. Combined, the costs of land and infrastructure usually exceed $1 million per mile. With bridges or other infrastructure, the costs increase significantly. Annual maintenance costs are estimated at $300,000 per mile. These economics challenge the development of railroads in many settings.

Railroad operators provide service to shippers by way of several different forms of equipment. The boxcar is the conventional form of transport on railroads for packaged freight.

However, most cargo that moves by rail is not contained on pallets or boxes. Quite often, railroads carry bulk cargos. Some cargos require enclosure to protect them from the elements; others (such as coal and gravel shipments) require no coverage and can travel in open-top cars or flatcars. Covered hopper cars can carry 263,000 pounds of grains in bulk. Meanwhile, tankcars can transport up to 34,500 gallons of liquid cargo. Figures 2-8 through 2-11 show a variety of rail equipment used to carry large volumes of freight.

Figure 2-8   Rail boxcar for containerized freight.

Figure 2-9   Small covered hopper car for grains.

Figure 2-10　Flatcar with timber.

Figure 2-11　Rail tankcar.

The Rail Market

The market for railroad services in the United States is segmented based on the revenues of the carriers. Large nationwide carriers earning operating revenues of $378.8 million are labeled as Class I carriers, according to the Association of American Railroads. Class I railroads operating in the United States include Burlington Northern Santa Fe, CSX Transportation, Norfolk Southern, Union Pacific–Southern Pacific, Grand Trunk Corporation, Kansas City Southern, and Soo Line. These major railroads actually are combinations of several disparate railroads. The railroad industry has experienced immense merger activity. Examples include the combination of the Burlington Northern and Santa Fe railroads in 1995, and the Union Pacific–Southern Pacific merger of 1996. Furthermore, Grand Trunk is a combination of Grand Trunk Western, Illinois Central, and Wisconsin Central. Merger activity has slowed in recent years as a result of greater scrutiny by the U.S. Department of Justice. Despite the mergers, the United States does not have a single railroad that can provide coast-to-coast service. Instead, western and eastern railroads maintain interline agreements and alliances that allow them to provide coast-to-coast connectivity.

Regional railroads operate on a smaller scale, with at least 350 miles of track and at least $40 million in operating revenue. These companies typically operate within one state or a few neighboring states. Smaller railroads, with revenues of less than $40 million, are regarded as local or "feeder" lines. Many formed as a result of deregulation brought forth by the Staggers Rail Act, acquiring rail lines that larger carriers abandoned. Still other railroads specialize as switching and terminal railroads.

A growing segment for most large rail operators is the intermodal market. Intermodal transportation involves the combination of two or more modes of transportation for a single shipment. Standard intermodal cargo containers can be transported via truck, rail, or ship. Railroads are a common component of intermodal transportation because they usually enjoy efficient portside access for international shipping and can reach cities very efficiently over long distances. Major railroads also maintain strategic relationships with trucking companies to provide the door-to-door convenience that trucking services offer.

Increasingly, rail carriers around the world are seeking arrangements that provide greater connectivity across national boundaries. The United States and Canada enjoy high levels of connectivity and extensive interlining, particularly with the Canadian Pacific and Canadian National railways owning significant stakes in U.S. carriers (the Soo Line and Grand Trunk, respectively). Rail transportation between the United States and Mexico has increased markedly since the Mexican railroad was privatized in 1995 and operates at six major border crossings. The EU is seeking similar arrangements in Europe to leverage the economies of scale benefits in a larger market.

Water Transportation

Transportation by water is perhaps most obvious on the high seas, where giant oceangoing vessels travel from coast to coast with enormous volumes of freight. Ships account for more than 90 percent of overseas international transport. However, transportation by water can also occur inland via lakes and rivers. All means represent significant components of transportation where navigable waterways present themselves. Unlike other forms of transportation, the viability of water transport is largely dictated by location. Put another way, origin and destination locations that are not located on navigable waterways will find the use of water a remote possibility. The next sections review ocean and inland water transportation.

Ocean Transportation

Ocean shipping is one of the oldest and most reliable ways of transporting goods over long distances. It also represents one of the most economical means, when one considers the tremendous volumes that a large ship can carry. Consider, for instance, the Triple-E Class container ships that began sailing the oceans in 2013. These ships have a capacity of up to 18,000 20-foot containers, or 20-foot equivalent units (TEUs). The ships are 1,312 feet in length and weigh 55,000 tons when empty. Expressed in product terms, a single Triple-E ship can carry 182 million iPads or 111 million pairs of shoes.[7] Maersk, the world's largest ocean shipping company, ordered 20 of these vessels for service between Europe and Asia, at a price tag of $185 million apiece. This is a far cry from the days of sailing ships, traversing the great oceans with a simple compass under the power of sea breezes.

So-called steamships are not the fastest mode of transportation, however. Ocean vessels travel at speeds ranging from 10 to 26 knots (or 11.5 to 30 miles per hour). Increasingly, ships are reducing their speeds to improve fuel efficiency. "Slow steaming" can dramatically reduce the energy required to move these massive vessels. The Triple-E ships are expected to average 16 knots (18.4 miles per hour). But speed is not ordinarily the highest priority for merchandise traveling by ocean shipping. Customers of ocean transport are instead seeking the lowest possible price and predictable, reliable delivery. Traversing the great oceans of the world can take weeks. Shippers (customers of shipping) fret over variability, however, for it impacts their ability to serve customers.

Ocean liner service is offered for scheduled routes among major seaports around the world. Large container carriers operate multiple ships on established routes for multiple sailings per week. Bookings are based on full-container-load (FCL) and less-than-container-load (LCL) bases. LCL loads are often booked by brokers and freight forwarders that consolidate the shipments of multiple customers into FCL volumes. Shipping containers are of five different lengths: 20, 40, 45, 48, or 53 feet, with the 45-foot container representing an international standard (as established by the International Organization for Standardization or ISO). The benefit of standard containers is the ability to stack them on ships and to ease the transfer of containers among ships, rail, and trucks to accommodate intermodal transportation. Container ships represent approximately 13 percent of the world's ocean fleet capacity.

Aside from container ships, a wide variety of other vessels commonly traverse the great spans of oceans. Bulk vessels carry a diverse assortment of commodities, including agricultural grains, coal, cement, and mineral ores. Bulk ships can have a single hold (cavity) for hauling a single commodity or multiple holds to transport different products. Similarly, tanker ships haul liquid commodities such as petroleum, gasoline, and chemicals in bulk quantities. Routes for bulk vessels are based on trade patterns, shipping from points of extraction to locations for further processing or consumption. Customers might elect to hire bulk carriers within the scope of a carrier's scheduled sailings or chartering vehicles for dedicated service. Private ownership and chartering is quite common among global producers and distributors of commodity materials. Bulk vessels represent approximately 17 percent of the world's ocean shipping fleet.

The balance of the world's ocean fleet consists of a combination of several different types of vessels and operators, including lighter-aboard-ship (LASH) carriers and roll-on/roll-off (RORO) carriers. LASH ships carry unpowered barges from one river port to another. They submerge so that barges can be loaded before regaining buoyancy for the transportation segment. Heavy-lift LASH ships are called upon for moving large infrastructure, such as oil rigs. They have also been used to transport crippled ships, such the *U.S.S. Cole* after it was attacked by terrorists in Yemen in 2000. RORO ships are designed to transport wheeled vehicles, hence, the name. They are used quite commonly for the transport of automobiles from manufacturing nations to import nations. The world's military also makes use of RORO ships to transport wheeled equipment. Passenger ferry ships often carry personal vehicles as well.

Inland Water Transportation

In addition to ocean transportation, water is an important means of moving cargo on inland transport segments. Transport over lakes and rivers can connect critical trade cities and provide access to ocean transport for global markets. Not all lakes and rivers are viable for commercial transport purposes, however.

In some cases, waterways must be widened, deepened, or straightened to permit large commercial vessels to be of any practical industrial use. When economically feasible, governments often develop infrastructure to overcome these impediments to inland water transportation. *Dredging* refers to efforts to deepen or widen a channel by excavating the river or lake bed to provide greater depth and draft for vessels to navigate. Rivers are susceptible to silt collections that can build up over time, which means that dredging can be a recurring need to ensure reliable transit in some channels.

Canals or trenches can also be developed to allow for short-distance connections between waterways. A canal is sometimes referred to as an artificial waterway because it involves manmade changes to the landscape to create navigable waterways. Canals have been in existence for centuries, dating back to ancient times in Egypt and China. Many canals require *locks and dams* to allow ships to travel from high-water to low-water areas (and vice versa) safely. A lock is a chamber that can fill with water to raise a vessel to a higher elevation or empty of water to lower the vessel to a low-water area.

One major infrastructural development of the modern era is the widening of the Panama Canal. This canal, which marks its centennial (100th year of operation) in 2014, provides a 48-mile-long shortcut between the Pacific Ocean and the Caribbean Sea. It is scheduled to have a new, larger series of locks in place and operational by 2015, allowing for higher volume and larger ships than the original canal can support.

Transportation by river is determined by several factors. The inland reach of the river is one consideration. In other words, does the river provide access to important trade locations, either sources of supply or points of demand? If so, is the river sufficiently deep and wide to accommodate vessels? Finally, what kind of vessels does the waterway support? Some rivers can support only small-volume, singular barge vessels. Barges are flat-bottomed vessels that might or might not have motorized forms of propulsion. Nonmotorized barges rely on external forms of propulsion, such as a tug or towboat. Small canal barges were once pulled by horses. Today's river barges are 195 feet in length, with cargo capacity of 1,500 tons. Liquid tank barges can be up to 300 feet in length, with capacity for a million gallons of liquid cargo, such as petroleum, fertilizer, or other chemicals.

For larger river corridors, such as on the Mississippi and Ohio Rivers in the United States, the Thames in England, and the Rhine in Germany, several barges can be linked in a single tow. In the lower reaches of the Mississippi River, it is possible to arrange up to 30 barges (5 barges wide and 6 barges long) in a single tow. Such multibarge tows provide for incredible efficiencies. Figure 2-12 illustrates a multibarge tow. In light of the considerable capacity of a single barge, aligning several barges in a single transport accommodates immense volumes. Figure 2-13 shows the comparisons with other conventional forms of transport. Transport by barge is not particularly fast, however. Depending on the speed of the river current, barges ordinarily travel downstream at speeds ranging from 8 to 14 miles per hour. The speed achieved against a current upstream is considerably slower, at 4 to 6 miles per hour.

The efficiency of river transportation bolsters the view of inland water transport as an environmentally friendly mode of transportation. Compared to other inland modes of transportation, it generates lower levels of hydrocarbons, carbon monoxide, and nitrogen oxide to move the same volume of cargo. For instance, a barge towboat generates approximately 0.0009 pounds of hydrocarbons to move 1 ton-mile, whereas a train generates 0.0046 pounds and trucks generate 0.0063 pounds, making them five to seven times higher in emissions.[8] Despite these arguments in favor of more environmentally friendly operations, one must also consider the potentially harmful effects of infrastructure development that alter the natural ecosystem. Dredging is one activity that has come under great scrutiny in recent years, and most developed nations employing commercial inland navigation require permits for any river development.

Figure 2-12   A multibarge tow.

Equivalent Units:
One barge = 15 jumbo hopper railcars = 58 large semis

One 15-barge tow = 2.25 100-car unit trains = 870 large semis

Equivalent Lengths:
One 15-barge tow = 0.25 miles
2.25 100-car unit trains = 2.75 miles
870 large semis = 11.5 miles (bumper to bumper)

Source: U.S. Department of Transportation

Figure 2-13   Comparison of barge capacity.

Another form of inland water transportation is provided by lakes. The Great Lakes system shared by the United States and Canada offers one such example. Connected to the Atlantic Ocean by way of the St. Lawrence River, the five Great Lakes (Lake Ontario, Lake

Erie, Lake Huron, Lake Michigan, and Lake Superior) provide extensive inland reach for oceangoing vessels, as well as connections among key industrial cities of the region. The Welland Canal provides an essential connection between Lake Ontario and Lake Erie, allowing ships to bypass Niagara Falls, although the canal's locks cannot accommodate large freighter ships (known as lakers). Freighters that exceed the 800-foot capacity of the locks on the Welland Canal travel among the 63 commercial ports on the four western-most lakes between Duluth, Minnesota, and Buffalo, New York. These ships can reach just beyond 1,000 feet (300 meters) in length. These ships primarily carry bulk quantities of minerals such as iron ore, limestone, coal, grain, sand, and gravel from extraction points to supply depots and processing sites. Figure 2-14 shows a large lake freighter.

Figure 2-14   A large lake freighter.

Given the northern climate experienced in these locations, the U.S. Coast Guard operates ice-breaking operations during the late autumn and early winter to keep the lakes navigable for lake ships. However, the cold of deep winter usually ceases these operations for two to three months each year. Severe drought conditions can also lower the water levels of the lakes, limiting navigation in summer months.

The Merchant Marine Act of 1920 (often referred to as the Jones Act), a form of cabotage regulation, requires that ships operating between origins and destinations in the United States be registered as U.S. carriers, operate U.S.-built ships, and employ crews of U.S. citizens. Although the Act was originally intended to apply to intracoastal water transportation, its provisions also find application across different modes of transportation in the United States.

Air Transportation

Air transportation is a common means of travel for passengers. It is less common for cargo, although people and cargo sometimes travel together. That is, many passenger airlines also operate cargo divisions, allowing freight to be transported in the lower cargo hold on scheduled passenger aircraft. A Boeing 777 airplane, for instance, can carry 263 passengers in the upper hold of the plane and about 9.5 tons of cargo in the lower hold (or "belly") of the plane. Transportation of mail and other forms of time-critical freight is quite common on these scheduled passenger routes.

Other air carriers specialize in the carriage of cargo, dedicating the full capacity of the aircraft to supporting cargo transport. All-cargo carriers can operate purely domestic routes or travel internationally. These carriers focus on long-distance routes and those over water among major trade cities. The value of goods transported by air cargo carriers is usually quite high, with the average value estimated at nearly $60,000 per ton. An example includes the transport of time-critical fashion merchandise from Hong Kong to retail centers around the world. The time-sensitive nature of fashion goods often demands that they arrive in retail stores as soon as possible to ensure that they meet the needs of a sophisticated market.

Another form of air carriage is the service provided by integrated carriers such as DHL, FedEx, TNT, and UPS. These carriers maintain an extensive fleet of aircraft for domestic and international services. Yet they also operate extensive surface transportation networks to support complete door-to-door services seamlessly across multiple modes. Integrated carriers are limited to a select few operators that have the immense capital resources to compete effectively on the ground and in the air.

Cargo aircraft can be quite expensive: The cost of a new Boeing 777 approaches $300 million. Despite the significant price of aircraft, the fixed costs associated with air operations can be lower than those of other forms, such as railroad and pipeline. The big difference here is that the infrastructure required for airlines includes the takeoff and landing facilities and the means of tracking aircraft (which, again, can be expensive). But often, many different carriers share these facilities. Very few air carriers maintain private airfields. Furthermore, the skies between origin and destination points are free. Railroads and pipelines, on the other hand, must lay infrastructure between the origin and destination, both owning the land and maintaining the infrastructure after it is built.

Air cargo is particularly sensitive to an imbalance in transportation movements. Whereas most passenger transportation tends to be bidirectional (people usually buy round-trip tickets), much freight flows from manufacturing centers to markets, yet the volumes flowing in the opposite direction are often much lower. Air carriers, therefore, price transit quite aggressively on the headhaul or outbound trip, with generous discounting on the return trip to entice volume.

Pipeline Transportation

Transportation by pipeline is among the least common and least known modes. Pipelines are ubiquitous, even though they are rarely seen. For companies that move massive volumes of fluid material over long distances, pipelines can be an integral component of supply chain operations. Whereas trucks, trains, ships, and planes are presented in clear view of the public, pipelines are often built underground in populated areas and typically run above ground only in remote areas. Yet pipelines serve as a primary means of transporting crude oil, refined oil, gasoline, and natural gas.

As Chapter 1 noted, the United States has more than 1.7 million miles of pipelines. More than 1.2 million miles are used to distribute natural gas. Much of the balance is dedicated to lines for crude oil and oil products. In the case of crude oil, pipelines called gathering lines collect the commodity in the oilfields. At a gathering station, large lines transport the crude oil to a refinery or processing facility, where the oil is converted into fuel. Product lines then distribute the refined product to storage locations near markets and major customers for final distribution.

Pipelines are owned and operated by federal governments in many nations. This is typically consistent in nations that have a significant ownership stake in the energy supply markets. Pipeline ownership in the United States and many other settings, however, rests with private companies. These companies are typically multibillion-dollar energy and chemical companies that can afford the significant investment in building and maintaining pipelines. Similar to railroads, pipeline operators must call upon government support in the form of eminent domain, in which the government requires compulsory forgiveness of private property to secure the right-of-way for the line. The provision of eminent domain is among the few government provisions that pipeline operators typically receive in the United States because it is the only mode that receives no form of direct subsidy from the federal government. The mode also is unique because it is regulated by the Federal Energy Regulatory Commission (FERC), an agency within the U.S. Department of Energy. The other modes of domestic ground transportation (interstate trucking and rail) are regulated by the Surface Transportation Board.

In addition to the line itself, pipelines require investment in pump stations, which provide the means of propulsion for the fluid material in pipelines. Large pump stations might have their own power plants colocated with them, particularly in remote locations. Small crews of 10 to 25 people can operate a typical pump station, making labor costs extremely low for a long pipeline. The Alyeska Pipeline, which runs a distance of 800 miles from the oil fields in Prudhoe Bay to Valdez, Alaska, requires only 450 workers.

Pipelines are generally regarded as extremely reliable yet slow. Liquid product typically flows through a pipeline at speeds of 3 to 5 miles per hour. Yet as long as the line remains intact and pump stations operate as expected, product can flow continuously. The diameter of the pipe determines the overall volume efficiency. A 36-inch pipe can carry 17 times the volume of a 12-inch pipe. So although it might cost 3.5 times more to construct the larger pipe, a pipe that size can generate a quick payback as long as the capacity is needed.

Despite a strong record of safe performance for operating pipelines, safety and environmental concerns remain. Environmentalists fear that the construction of the lines can impact sensitive ecosystems. It was feared, for instance, that placing the 48-inch line above the ground in the trans-Alaska pipeline would interfere with the migrations of caribou and other wildlife. To counter this concern, some sections of the pipeline are buried in the ground or elevated to allow animal and human traffic to occur unfettered. However, burying and elevating lines costs considerably more. Furthermore, concerns remain that breaks in buried lines will be difficult to detect and repair. Areas prone to earthquakes and seismic activity require that lines be built with a degree of flexibility to accommodate tremors and small quakes, again adding to the expense.

Intermodal Transportation

In 1956, a major innovation in transportation occurred in Baltimore, Maryland. Entrepreneur Malcolm McLean improvised a new form of transportation by converting a retired World War II oil tanker shipper into the first lift-on/lift-off container ship, called the Ideal-X. The premise involved not only the conversion of the ship, but also the creation of a standard cargo container that could be moved by flatbed trailer truck and easily transferred to a ship. Building standard-size containers meant they could be placed optimally on the deck of the ship, and they eventually were designed so that they could be stacked several layers high for increased efficiency. This innovation marked the beginning of modern intermodal transportation.

Intermodal transportation involves the use of two or more modes of transportation to move a shipment from origin to destination. Intermodal transportation leverages the relative strengths of each mode or overcomes a challenge faced by a single mode. For instance, the Ideal-X allowed truckload-size shipments to move quickly from an inland origin to a shipping port. After it was transferred (or "transloaded") to the ocean ship, the shipment benefited from the volume and distance efficiencies the ship offered. Finally, upon arrival at the shipping port near the customer, the freight benefited from the convenience of door-to-door delivery that trucking afforded. The significant innovation here was that after the goods were loaded in the container, they need not be handled again until delivery. That is, there was no requirement of handling loose cargo or bags of goods on a repeated basis, as had been the convention from the time of Viking ships! Over time, the standard intermodal shipping containers could move by rail as well as truck and ship. The introduction of intermodal rail improved the efficiency of container transport over long distances on the ground. It is possible on some rail corridors to double-stack the containers to enjoy even greater efficiencies.

Intermodal transportation often involves road, rail, and water transportation in succession. In the first segment, a container is collected at a shipper location by truck. The truck delivers the container to an intermodal rail yard, where it is transloaded to rail for long-distance ground movement to an export shipping port. The container can then ship along with thousands of other containers on the ship. At the import port location, the container is transferred to the ground for movement by rail or truck to the customer location. The development of international standards for container sizes and configurations supports this seamless transition across different modes in multiple countries. In 2012, 12.3 million intermodal shipments moved in the United States, with 80 percent of those moves occurring in intermodal containers. Figures 2-15 through 2-20 show some of the different means of intermodal transportation today.

Figure 2-15 Intermodal shipping container.

Figure 2-16   Trucks deliver containers to port.

Figure 2-17   Gantries and cranes transload containers at ports.

Figure 2-18   Intermodal ocean container ship.

Figure 2-19   Intermodal containers aboard a large river ship.

Figure 2-20   Intermodal rail transportation.

Innovations continue to be introduced in intermodal transportation, including the size of containers and the ships to carry them (both of which are increasing). The Triple-E fleet of ships that Maersk launched beginning in 2012 has a capacity of 18,000 20-foot equivalent units (TEUs), or 9,000 40-foot equivalent units (FEUs). Many shipping ports will be required to dredge deeper sailing channels and develop larger quays with more dock space and conveyance equipment to accommodate these behemoth vessels.

Note that air transportation is also involved in intermodal transportation with the use of standard air containers employed by integrated carriers such as UPS, FedEx, and DHL. These containers are designed to fit in the trucks of these carriers, as well as in their cargo aircraft, with fast and easy transfer of the containers at airports. The containers are moved from truck to airplane using roller conveyance. Even the cargo hold of the aircraft is equipped with roller floors to allow the cargo to flow easily into the proper position in the aircraft and back off again. Figures 2-21 and 2-22 show air operations using the containers and roller equipment.

Figure 2-21   Air containers.

Figure 2-22   A "belly" air container.

Pipeline is the only mode, to date, that does not find involvement in intermodal transportation. The fluid materials that are transported via pipeline are often moved by other modes upon induction into the pipeline. Fluids also are commonly delivered via other modes after moving by pipeline to a distribution point. Consider oil extracted in the north shore of Alaska, transported to the south of Alaska via the Alyeska Pipeline. From Valdez, it is transferred to an oil tanker ship for transport to the U.S. Northwest. From here, the product might move by rail or truck to customers. Yet the distinction of intermodal transportation is the ease of conveyance afforded by using a single container to move a consolidated volume of goods among the various modes of a shipment. This does not currently exist in pipeline transportation.

Interesting designs exist, however, for transporting nonfluid materials in pods via pipeline. Imagine the pneumatic vacuum tubes that drive-in banks use to transfer currency and small documents between a teller and consumer from the comfort of your car—but on a much larger scale. Extensive studies have examined the viability of such forms of transportation. Designs for pneumatic capsule pipelines (PCPs) are under consideration in Europe and Asia for the transfer of much larger containers for cargo. Researchers in China claim to be working on designs that will transport people and goods at speeds in excess of 1,000 kilometers per hour (faster than today's commercial jet aircraft!) within 10 years. It is conceivable that these pods could be transported by other modes of transportation to provide point-to-point delivery on a more comprehensive basis, hence qualifying such a move as intermodal.

Summary

Key takeaways from this chapter include:

- Five basic modes of transportation exist: road, rail, water, air, and pipeline.
- Each mode has its relative strengths in terms of service and cost.
- Most products are transported by road at some point in their distribution.
- Nations with high-performing railroads and navigable waterways have advantages in moving large volumes of freight over long distances very efficiently.
- Pipelines are expensive to build and maintain, but they operate extremely efficiently in the transport of fluid materials.
- Intermodal transportation leverages the advantages of two or more modes of transportation to support a shipment's movement.

Endnotes

1. The term *forwarders* refers to transportation service providers that work across modes.

2. In the United States, federal guidelines permit truck-trailer combinations to weigh up to 80,000 pounds. The truck and empty trailer can weigh up to 36,000 pounds, leaving approximately 44,000 pounds available for freight. Individual states allow higher or lower weight restrictions, because matters of truck safety are regulated at the state level.

3. A simple Google search for "Ohio trucking services" yielded 2.14 million results (as of September 8, 2013). Although the number of trucking companies offering service in the Ohio market is far less than this, it speaks to the ease with which available carriers can be found.

4. States continue to regulate matters of safety and social concern, which often has implications for the economic vitality of motor carriers.

5. The Owner-Operator Independent Driver Association (OOIDA), a trade association representing small operators, boasts more than 160,000 members in the United States.

6. Source: CSX Transportation Web site, at www.csx.com/index.cfm/about-csx/projects-and-partnerships/fuel-efficiency/.

7. John Francis Peters (2013), "Building the World's Biggest Boat," *Bloomberg Businessweek* (September 9): 44–50.

8. C. Jake Haulk (1998), *Inland Waterways as Vital National Infrastructure: Refuting "Corporate Welfare" Attacks,* Allegheny Institute for Public Policy.

3

THE ECONOMICS OF TRANSPORTATION

Transportation economics occupies its own branch of the economics discipline. Economics is concerned with determining the best means of allocating scarce resources to achieve the greatest overall benefit. Governments use economics to determine projects in which they should invest, such as those of transportation infrastructure (for roadways, bridges, ports, locks and dams, among others). Businesses also must use economic principles to make the most informed and educated decisions in resource allocation. On matters of transportation, these decisions include determining how many facilities in which to operate for serving customers (distribution network design), whether to invest in private fleet operations or buy services from carriers in the marketplace, and which modes to use for inbound and outbound shipping, just to name a few.

This chapter focuses on the economics of transportation operations for *carriers*, firms that provide transportation services. With an understanding of the costs incurred by carriers, we examine the implications of pricing services for customers. Under efficient market conditions, carriers should earn sufficient revenues to cover costs, yet competition in the market will prevent carriers from earning exorbitant or excessive rents. This example is initiated by reviewing the different ways to view costs that carriers face in their operations.

Accounting Costs and Economic Costs

One way to distinguish the costs of transportation operations is to consider *accounting costs* and *economics costs*. Accounting costs involve actual outlays in money and are recorded in the financial statements of a company. Economic costs, on the other hand, reflect the revenue that is lost in pursuing a different course of action. Consider, for instance, that a carrier elects to invest in information technology rather than a new truck terminal. The investment in the information technology represents an accounting cost. Economic cost is found in the incremental revenues that the company misses by *not* investing in the expanded capacity provided by a truck terminal. Economic costs are often called *opportunity costs*. The determination of opportunity costs is often speculative because it reflects an action not taken. Despite these concerns, it is essential to estimate this to decide how to invest scarce funds.

Fixed Costs and Variable Costs

Another way to distinguish the costs carriers incur is to distinguish accounting costs as either *fixed costs* or *variable costs*. Fixed costs refer to expenses that are incurred regardless of the level of activity. Variable costs increase with more activity and decrease with less activity. Succinctly, infrastructure, facility, and vehicle costs tend to be fixed, whereas costs associated with the operations of movement are typically variable. Fixed costs are incurred whether or not a shipment moves. However, fixed costs do experience step functions that reflect demands of assets. When shipping volumes exceed the capacity of a fixed asset, the carrier might invest in more capacity, incurring additional fixed costs. In this sense, fixed costs can vary with activity. New assets are acquired to accommodate capacity shortages, or existing assets are sold to deal with overcapacity.

Variable costs are accrued in different ways. Distance is one important dimension of variable cost. The longer the distance for a shipment, the greater the cost, and vice versa (shorter distance translates to lower cost). Distance affects cost through the consumption of fuel and driver time (wages). Volume is another dimension of variable costs. The volume of a shipment can affect cost in two ways: 1) More volume requires more material handling of freight (in loading and unloading freight), and 2) more volume translates into greater weight and a higher consumption rate for fuel to move the heavier load.

The summation of fixed and variable costs amounts to *total cost*. Figure 3-1 illustrates typical behavior for fixed, variable, and total costs in the case of distance. Note that fixed cost is a flat line that does not change with distance. Variable cost, however, rises with distance. The total cost curve, then, is the sum of fixed cost and variable cost, and increases at the same rate as variable cost. Discussions of variable costs lead to an understanding of *marginal cost*. Marginal cost refers to the additional cost incurred with an incremental increase in an activity.

In Figure 3-1, marginal costs are reflected in the slope of the total cost curve (and variable cost curve, which is the same). Close examination of the total cost curve suggests that the slope of the curve decreases somewhat as distance increases. This decrease in marginal cost suggests that an economy of scale is present. An *economy of scale* is found when, the further one pursues an activity, the less each additional unit costs than the one before it. In our example with distance as the focus, each additional mile costs less than the one before it. This is called the *tapering principle*. Economy of scale is also found with volume, as suggested by our discussion of how variable costs can increase.

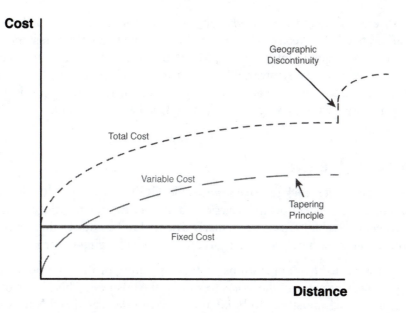

Figure 3-1 Fixed, variable, and total cost.

The total cost curve reflects jumps (sudden increases) from time to time, as in Figure 3-1. In the case of distance, at some point, a shipment can reach limits beyond which it can no longer support a shipment, such as when a truck runs out of road in a transoceanic shipment. The truck must hand off the load to a ship or airplane. This is called *geographic discontinuity*. Therefore, the total cost curve reflects the cost of the movement by truck, and then when the truck reaches its terminal point, the total cost curve takes a jump to a higher level before reflecting the marginal cost of the second mode of transportation employed in the shipment (either air or ship). Of course, when the airplane or ship hands off the shipment to another mode for final delivery to the customer, we would see another jump.

Similarly, sudden increases occur in the total cost curve for shipment volume. With increasing volume, a shipment exceeds the capacity of a vehicle to carry the load at some point. This is true of the weight capacity and the space capacity of a vehicle. When a vehicle reaches its maximum weight, it is said to have "weighed out." When a vehicle reaches its spatial capacity, it is said to have "cubed out." To ship volumes in excess of a vehicle's weight or a container's weight or cubic capacity, we must enlist another vehicle (or container). The enlistment of the second vehicle would result in a jump in the total cost curve, reflecting the fixed-cost investment in the second vehicle (or container).

Some costs are considered *semivariable* or *semifixed*, not distinctly variable or fixed. As noted earlier, fixed costs technically remain "fixed" only to a point; then another truck, terminal, or other equipment is needed. Maintenance cost for a transportation vehicle is sometimes considered a semivariable cost, or mixed cost, because it tends to increase with the level of transportation activity, but not necessarily in direct proportion to the change in activity. Overtime wages are often regarded as semivariable.

Carrier Cost Metrics

Transportation carriers seek to price service at levels that ensure profitability. Revenues must exceed costs to yield profits. Carriers in all modes seek to utilize their assets as productively as possible. This is particularly essential in businesses that serve intensely competitive markets where profit margins are likely challenging to earn.

Where profits earned from operations (that is, operating profits) are possible, carriers must aggressively reinvest in the assets to sustain the business. This is true of railroads and pipelines, which must pay to build and maintain dedicated rights of way (as discussed in Chapter 2, "A Survey of Transportation Modes"). The measurement used to reflect the efficiency of a carrier's business is *operating ratio* (OR). The calculation of OR follows:

Operating ratio = [(Operating costs) / (Operating revenues)] × 100

The lower the OR, the more money a carrier has available to invest back into the business. The OR serves as a valuable measurement of a carrier's financial health. Although it is less meaningful when comparing carriers in different modes, the OR can be useful in assessing the operational efficiencies of carriers competing in the same mode. Table 3-1 lists the OR of several carriers competing in the less-than-truckload (LTL) market in the United States. The average OR of 97.0 percent for 2013 indicates that the carrier margins in this sector are quite thin. Again, the lower the OR value, the more funds a carrier has to reinvest in the business.

The OR serves as an important metric for both the carrier and anyone who might evaluate the carrier, including current and prospective customers, investment analysts, financial institutions, and, where applicable, regulators charged with ensuring the economic financial fitness of a carrier. Carriers in financial distress are feared to be at greater risk of safety violations.

Table 3-1 Operating Ratios for the U.S. Less-than-Truckload (LTL) Sector

For Quarter Ending March 31, 2013				Data in $ Thousands		
Carrier	YRC Worldwide	Arkansas Best / ABF*	Old Dominion	Conway**	Saia***	Total Carriers
Total Operating, Revenue Including Fuel	$1,162,500	$520,687	$532,575	$827,736	$273,795	$3,043,298
Change 2013 from 2012	-2.70%	18.10%	7.10%	-0.40%	-3.60%	2.70%
LTL Tonnage	-4%	4.20%	3.50%	-1.30%	-1.60%	
Net Income	-$24,500	-$13,395	$40,533	$16,024	$9,155	$27,837
Change 2013 from 2012	Had loss of $79.8 million in Q1 2012	Had loss of $18.16 million in Q1 2012	30.40%	-53.60%	65.40%	
Net Income as a % of Revenue	-2.10%	-2.60%	7.60%	1.90%	3.30%	1.60%
Net Income as a % of Revenue 2012	-6.70%	-4.10%	6.30%	4.20%	2.10%	0.30%
LTL Operating Ratio 2013	99.10%	105.50%	87.60%	98.10%	94.70%	97%
LTL Operating Ratio 2012	104.10%	105.50%	89.10%	95.80%	95.90%	98.10%

* Includes data from its Panther Express Unit, except OR percent.

** Conway numbers refer only to its LTL group, not the business as a whole, which includes Menlo Logistics, a truckload business, and other units. Conway Income refers to operating income only for LTL group, before other expenses that would be included in full net income number, as it is posted for the other carriers.

*** Saia's revenue numbers include a truckload unit that represents about 15% of total tonnage; revenues are not broken out.

Another financial metric of significant interest to carriers is *cost to serve* (CTS). CTS is the carrier's estimate of the specific costs incurred in providing service to a distinct customer. This is critical input for pricing the carrier's services appropriately. Charging the same price for service offered to different customers might be simple, but it is laden in hazards for the carrier. Relying on average costing assumes that customers are uniformly consuming the carrier's resources, which is a virtual impossibility. It is a mistake, for instance, to assume that any mile or any unit of distance is the same as any other. Whereas a mile on an open highway might take only 1 minute to complete, a mile in a congested downtown during rush hour might take 20 minutes, consuming considerably more of the driver's time and fuel. Similarly, recall from Chapter 2 the difference in speed for a barge traveling downstream with the river current, as opposed to upstream against the current (12 miles per hour versus 5 miles per hour). The towboat supporting the barge would consume more fuel heading upstream than downstream. Also, consider the customer that requires service provisions such as special handling of freight or unique communications and support. These simple examples underscore the importance of understanding a customer's specific demands and the costs incurred to meet those demands, or the CTS.

One method that many organizations have applied over the years to ascertain the cost to serve a distinct customer is *activity-based costing* (ABC). Devised in the early 1980s, ABC became common among infrastructures for assigning overhead costs to products and customers. Transportation and logistics companies have employed ABC as a way to understand the relationship between customer demands and resource consumption. As its name implies, the method requires a company to track its activities. This is accomplished with mapping processes, such as loading and transporting freight, and administrative processes, such as communicating with the customer and issuing the invoices. After the processes are mapped, they are linked to the resources consumed and costs incurred in performing the processes. This requires examining the company's financial records and expenses, including general ledger activity, statements of cash flow, and income statement. Today's CTS systems can ease this step. Out of this, a cost per activity is devised for each of the various activities the company performs in serving costs. The final step, then, is to integrate the activity log in the analysis. This illustrates how frequently the service provider performs each activity for a customer over the allotted time period (usually the previous 12 months). The company sums the activity costs for each customer to determine the CTS. Figures 3-2 and 3-3 illustrate the logic employed.

In summary, ABC provides a more accurate way to capture costs for distinct customers than relying on averages, a method that assumes that customers are uniform. The ABC method is not easy to establish or to maintain in complex businesses, however. One important yet often overlooked opportunity for simplifying the analysis (and also for enhancing decision making) is to include only the costs that would go away if the customer was no longer a customer. In other words, include only the operational and administrative costs that would increase with more activity directed to a customer or decrease with less activity. In other words, if a carrier would not close a terminal or maintenance facility if it no longer served that customer, there is no reason to attempt to allocate these costs to this customer, or any other. This one simplification can make the ABC analysis much easier and more insightful for informed decision making, such as pricing averages.

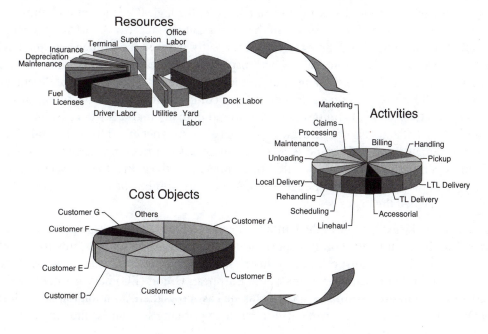

Figure 3-2   Translating resources in activity-based costs: the cost flow.

The calculations

Resources ▶	$\dfrac{\text{Cost}}{\text{Driver}}$	X	$\dfrac{\text{Driver activity}}{\text{Cost object}}$	=	$\dfrac{\text{Cost}}{\text{Cost object}}$
Cost data from the general ledger or income statement	Overall driver activity level in records, observation & employee estimates		Documentation of cost object activity in customer/sales records, observation & employee estimates		Total service cost for a cost object (customer, service, product)

The data sources required for each calculation

Figure 3-3   Translating resources in activity-based costs: the calculations.

Carrier Pricing and Costs for Shippers

Transportation prices are referred to as *rates*. Several factors influence the rates that carriers charge for their services. In many nations, governments determine the prices carriers can charge. Regulated markets are intended to ensure sufficient availability of services for shipping customers and ample margins for participating carriers, many of which are state-owned in heavily regulated environments. The United States maintained pricing regulations for interstate trucking and rail services through the Interstate Commerce Commission (ICC) until the passages of the Motor Carrier Act (MCA-80) and the Staggers Rail Act in 1980. These deregulatory acts greatly reduced the influence of the ICC and rate bureaus on economic matters in transportation. Research indicates that markets for trucking and rail service changed dramatically as a result of these acts of deregulation, benefitting many shippers. In light of the competition induced by deregulation, services greatly expanded and prices were lowered in many markets where volume and competition dictated. It is not accurate to say that all parties benefited from deregulation, however: Small shippers in remote areas saw increases in prices and declining service offerings. Similarly, carriers that could not compete effectively in a free market lost the protections they enjoyed under regulation. Yet in the opinion of most observers, the gains outweighed the losses.

As a result of the experience in the United States, several governments around the world are relinquishing influence on transportation markets. In some settings, this translates into privatization of transportation industries, including air, rail, water, and pipeline transportation. For instance, nations of the European Union (EU) have greatly liberalized air transportation within each nation as well as across the EU nations since the late 1990s. In other settings, governments are removing pricing standards, allowing greater freedoms for pricing.

When market forces are allowed to freely determine rates, a primary input is the cost the carrier incurs in providing the service, as illustrated thus far in this chapter. Carriers should seek to cover their costs, at a minimum, to ensure profitability. Several inputs determine the cost of service. A primary input is the cost of physically collecting and moving the shipment over a specified distance. The carrier incurs the operating expense of positioning the equipment for the pickup, performing the move, and returning the equipment to its domicile location or its next pickup location. However, beyond the operating expenses of driver wages and benefits, fuel, and vehicle depreciation, a host of additional costs are incurred, including the administrative activities associated with planning the shipment, dispatching equipment, tracking the shipment, and invoicing for the service. Yet more additional costs are incurred in promoting and selling the service to the prospective customer. Still more costs are associated with paying insurance premiums and securing licenses and registrations. Clearly, the prospect of covering costs is a challenging proposition in and of itself for carriers, and especially so in highly competitive markets.

Where competition is high, rates will be priced aggressively low among carriers that are vying for market share. When competition is lessened, however, rates will tend to be higher. Competitive markets keep margins very tight in many segments of transportation. As an example, the dry van truckload market in the United States might be described as hypercompetitive. No single carrier has a demonstrable share, and the market is highly disaggregated. This business is often regarded as "pennies on the mile" business in light of the competitive rates and very narrow margins enjoyed in this highly competitive segment. Carriers in these situations are always looking for opportunities to gain healthy margins in less competitive routes or lanes to counter the intense price competition they face in head-to-head rivalries.

The countervailing force to the carriers' cost is the customer's willingness to pay for service. A customer's willingness to pay can be based on different factors, including the comparison prices other carriers offer for similar services and the cost that the shipper would incur to perform the service itself. When transportation is essential and no viable alternatives are available, prices tend to be higher. When customers are willing to pay the higher price for service, demand is said to be *inelastic*. *Elastic demand*, on the other hand, is associated with changes in demand associated with changes in price. When price increases, demand decreases. With *unit elasticity*, a specific percent increase in price results in a decrease in demand of the same degree (that is, the same percentage). In cases of imperfect market conditions with inelastic demand, government involvement is sometimes employed when a carrier charges exorbitant prices. Conversely, shippers have also been known to exert inordinate influence on transportation markets. Perceived abuses among rail carriers in the late nineteenth century led to the formation of the ICC.

Carriers take factors other than cost into consideration when they can freely price transportation services. The potential risk associated with a shipment is one of these factors. Risk can be associated with the potential for loss or damage of the goods. High-value goods carry a high level of financial liability for the carrier. Therefore, the carrier might charge more to accept this risk. Other goods are particularly prone to loss or damage. Perishable fruits and glass products are examples of such items. Hazardous materials and goods that require special handling or security can also see higher rates. Special services, such as temperature or climate control, also command a premium in rates. Temperature control requires additional energy and additional risk associated with a product that can be ruined when exposed to excessive heat or cold.

Another factor influencing price is the *balance of freight flows*. If a carrier can expect to earn revenues on the return (backhaul) trip as well as the outbound (fronthaul or headhaul) trip, the carrier will be more willing to offer the fronthaul trip to a shipper at a lower price. However, if revenue is possible only on the fronthaul, the carrier will try to cover its costs of operations for both the outbound and return trips with the one shipment. In fact, it behooves the shipper to help the carrier find backhaul loads, to reduce the burden of covering costs for both trips. The shipper itself might offer backhauls as one way to ensure carrier balance and better pricing.

Freight imbalances might be temporary or long-term challenges for carriers. Consider, for instance, the seasonality of agricultural harvests in late summer when the crops of an entire region might be transported within a time period of a few weeks. While loaded trucks, trains, and barges might ship the harvested grains from the farms and grain elevators to processors, there is likely to be little return freight. Farmers thus are prepared for this premium or elect to perform the transportation services themselves. A longer-term imbalance is found in shipping goods to south Florida. The state of Florida is a "net consumer" state, meaning that it consumes a larger share of manufactured goods than it produces. This translates into hundreds of thousands of trucks delivering the products to serve this southern region in a peninsular state. Yet few trucks are able to find freight for revenues on the return trip to the manufacturing states in the north, resulting in higher freight rates to deliver goods to south Florida. An exception occurs, though, when fruits and vegetables are harvested in the region for distribution in the north. During this time, trucks can enter and leave Florida with freight, allowing the carrier to earn revenues on both the fronthaul and backhaul routes.

Expressions of Transportation Rates

Transportation prices can be expressed in different ways. The simplest expression is a flat rate between two points. Carriers operating over fixed routes, such as railroad carriers, often charge a specific amount between two points in the network. This is called a *commodity rate*. Pipelines and water carriers might also charge a fixed rate for service between two fixed points in the network. In many cases, the volume of freight does not factor into these fixed point-to-point rates.

Trucking companies with greater variation in routes might instead choose to price service based on distance. Charging on a per-mile basis, for instance, allows a carrier to ensure coverage of variable costs on nonfixed routes. Truckload carriers often express rates on a per-mile basis. Additional charges, called surcharges, might be added to these rates to accommodate inflation in fuel prices or to accommodate the added time of city deliveries.

Another form of pricing is found in *class rates*. Class rates are intended to simplify multiple criteria that might factor into pricing. LTL trucking companies are the most likely carriers to employ class rates. Whereas truckload carriers are inclined to charge flat point-to-point prices or per-mile prices for service, LTL carriers (and those with opportunity costs associated with empty space in the transportation vehicle) charge on a volume basis.[1] Recall that truckload carriers place limits only on the maximum weight of the load and charge the same price regardless of the volume. LTL carriers, on the other hand, seek to generate revenue from several different shippers to fill their trailers. When a trailer has empty space or unused weight capacity, this represents a lost opportunity for earning more revenue. For this reason, LTL carriers embrace a tradition of charging customers for the amount of the weight or space capacity that the customer's load will consume in the vehicle.

To accommodate these multiple dimensions, class rates are expressed in terms of weight and commodity classes. In the United States, the rates are expressed in dollars per 100-pound weight increments (also known as "century weight" or "hundredweight"). The commodity class, then, reflects *density* of the goods being shipped. Density is the measure of weight, given space. An item that is very dense is quite heavy, given little consumption of cubic space in the vehicle. Metals and liquids are examples of dense products. A product with low density consumes little weight capacity, given its space consumption. Because class rates are expressed in dollars per hundredweight ($/cwt), a product with low density will have a higher classification, reflecting a higher rate per hundredweight since it will consume "more than its fair share" of the space capacity of the vehicle. Consider a shipment of chairs. Even though a single chair might weigh 20 pounds, a truck would have a lot of open space even if the chairs were stacked and nested. In other words, the truck would be shipping not only chairs, but a lot of "air" because the chairs leave many open gaps, even when nested. Therefore, chairs would have a higher classification than denser products, yielding a higher price for the shipment of these items via LTL carriers who are concerned with how their space and weight capacity are used in their transportation vehicles.

Aside from density, additional factors considered in the determination of commodity classes include the stowability, handling, value, hazardous nature, and liability associated with handling the product in question. As products change in design or are in any way adapted, the classification assigned to a product can change. This is accommodated by way of deliberations among a committee of interested stakeholders consisting of both shippers and carriers. In the United States, the National Motor Freight Traffic Association (NMFTA; www.nmfta.org) is the party responsible for maintaining the commodity classifications. The NMFTA maintains the National Motor Freight Classification (NMFC), which codifies products in 18 different classes, ranging from 50 to 500. Products rated at Class 50 enjoy the lowest rates per hundredweight, as a reflection of their ease in "transportability." They are likely to be dense, easily stowed and handled, sturdy, and relatively free of hazard and liability. Materials such as bricks or packaged sand shipped on pallets are examples of Class 50 freight. At the opposite extreme, products rated as Class 500 are low in density, difficult to stow and handle, fragile, extremely valuable, and/or laden with hazard or liability. Gold and Ping-Pong balls are examples of Class 500 freight. Gold is extremely valuable, earning the high rating, and Ping-Pong balls are extremely light. Both products would receive a higher price (as expressed in $/cwt) than items with lower classifications.

The classification system was essential in a regulated trucking environment, and it remains intact today as a vital reference in the industry for simplifying negotiations in transportation pricing. Another simplification that remains is the use of *freight-all-kinds (FAK) rates*. An FAK rate is a negotiated rate that applies to one or more shipments of products with different classifications. For illustration, consider a shipment of 100 pounds of Class 50 product, 600 pounds of Class 100 product, and 1,300 pounds of Class 300 product. The carrier could charge different prices for the three respective product classes. However, the shipper might seek to negotiate a single class to apply to all three types of freight in the shipment. The shipper would be wise to seek a lower FAK classification to enjoy a lower rate for the entire 2,000-pound shipment. The carrier will counter with a recommendation of a higher FAK classification. If the shipper and carrier agree, the FAK rate will appear on the freight bill for the shipment with the determined class and associated rate applying to the full load. Shippers employ FAK rates as a way to negotiate lower rates than they would enjoy under conventional class rating. The strategy offered simplification for the carrier in the days of manual invoicing, but the practice remains despite advanced automation used in issuing and auditing freight bills.

Airlines use a similar basis for calculating transportation costs as LTL carriers. This is because airlines, similar to LTL carriers, are concerned with how the space and weight capacity of the vehicle (aircraft) is used among a multitude of different customers. When an opportunity cost is associated with how the space or capacity is used, variable pricing is used. The airlines employ a density measure called *dimensional weight*. This is the standard weight for the cubic dimensions of a load. A common standard air carriers use is 10.4 pounds per cubic foot—or, inversely, 166 cubic inches per pound. For shipments employing the metric system, 200 kilograms per cubic meter (5,000 cubic centimeters per kilogram) and 166 kilograms per cubic meter (6,000 cubic centimeters per kilogram) are often employed as standards. Different carriers employ different dimensional weight standards, and the standards can vary between purely domestic and international transportation.

In pricing air cargo service, the actual weight of the air cargo shipment is compared to the dimensional weight standard the carrier offers. The shipper is charged based on the heavier of the two. Suppose a carrier uses a standard density of 10.4 pounds per cubic foot to price freight. What if a customer seeks to ship a load that measures 4.5 feet in length, 4 feet in width, and 3.75 feet high, and weighs 280 pounds? The cubic dimensions of the load are (4.5 × 4 × 3.75): 67.5 cubic feet. Dividing the weight (280 pounds) by the load dimensions (67.5 cubic feet) yields a density of 4.148 pounds per cubic foot. This is less than the standard density of 10.4 pounds per cubic foot. Therefore, the carrier will use the standard density (instead of the actual weight) when charging for the load. Dimensional weight pricing is said to penalize shippers of low-density freight. However, to be fair, the pricing reflects the operational realities of the air carriers because they have limited capacity in terms of weight and space in which to maximize revenues and cover the high costs of operating in an expensive mode of transport.

A final factor to consider in determining rates for basic transportation service is a *minimum charge*. Despite the formulas applied to determine prices, carriers usually ensure that a threshold price or minimum charge applies to all shipments, regardless of the shipment's distance or volume. The carrier incurs operating costs, administrative costs, and opportunity costs with all its commitments. Therefore, the carrier must ensure that all loads have at least a prospect of profitability after covering expenses. The minimum charge carriers offer will vary considerably, depending on the mode and resources they deploy in providing service.

Additional Services and Fees

The price of the basic transportation service is called the *linehaul price*. This embodies the transit of products from origin to destination. However, carriers might incorporate additional fees into their pricing strategies. These include *stop-off fees* for dropping off freight or picking up freight at intermediate points aside from the origin and destination of an original shipment. The fee is justified in light of the additional time required to stop and accommodate the additional load. It is common for a shipper to hire a truckload carrier to ship multiple loads in a single shipment. One load is destined for an intermediate point, with another dedicated to the final delivery location. Carriers often accept these additional loads as long as the intermediate stop does not take the shipment too far "out of route" for the original delivery. The farther the distance is out-of-route, the more the carrier will charge for the stop-off. In addition, the carrier is likely to institute a provision for an extended time or wait associated with the stop-off that impairs its ability to meet the delivery requirements at the final destination. That is, any delays at the stop-off location imperil the commitment to the final destination.[2]

Carriers might also charge a fee for diverting a shipment from its original destination in favor of another location. *Diversion* or *reconsignment* applies when a shipper and carrier have agreed on the movement, but the shipper issues a change in destination, perhaps when the shipment is already in transit. Such an arrangement is most likely to occur in the road transport mode, which can more easily accommodate a change in the destination for a shipment. This might be the case for a shipment that was originally planned for one distribution center in a company's logistics network but that must be diverted to another facility deemed to be in a shortage situation. The carrier's reconsignment fee ensures that it is compensated for the inconvenience of repositioning its equipment and driver to accommodate the dynamic routing of the shipment.

A fee that is common to issue to shippers and receivers of freight is a penalty for holding up a transportation vehicle and, sometimes, the operator of the vehicle. *Detention fees* refer to delays in loading or unloading a truck, preventing the truck from being used for transportation services. Often not only the equipment is left waiting, but the truck driver is as well. Carriers typically offer a grace period of up to 2 hours for a shipper to load freight and a receiver to unload freight. When the service time exceeds this standard, the carrier might issue a fee for detaining the equipment and driver. This is a major concern in many nations where carriers are facing driver shortages. Germany, for instance, is facing a shortage that is expected to grow to 150,000 drivers by 2023.[3]

One reason for the driver shortage is the frustration of drivers who are forced to wait at pickup and drop-off locations. The tradition in driver wages is to pay the driver for distance traveled, but not for waiting. Under this scenario, drivers experience considerable aggravation, and many choose to leave the industry. A study conducted among truck-load carriers in the late 1990s found that the average longhaul truck driver wasted 33.5 hours each week waiting to load and unload freight![4] Adjustments in hours-of-service (HOS) that challenge drivers' available drive time place an additional premium on how the driver's time is spent. To combat the problem, carriers have become quite aggressive in charging detention fees and passing along much of the proceeds to the drivers. Shippers that refuse to pay the detention fees then run the risk of being declined service in the future. Furthermore, shippers that gain a reputation of detaining drivers and not paying detention fees gain a reputation in the industry as being uncooperative. The term *demurrage* applies to the fee that charter ship owners and rail carriers charge customers for extended use of vehicles and containers in their respective modes. Demurrage is usually charged on a *per diem* (daily) basis instead of being levied hourly, as for trucking.

Motor carriers sometimes enter into agreements that allow shippers several days to load or unload a trailer. *Drop trailer service* refers to a provision of leaving trailers with a shipper or receiver for an extended time to allow more loading or unloading at times that are convenient for the customer. Carriers charge a daily fee for such provisions. The provision of dropping trailers also allows optimal utilization of the driver's time.

Surcharges refer to any additional fee applied to a shipment. Fuel surcharges are among the most common. When fuel prices prove volatile, carriers incorporate a surcharge to protect themselves against the risk of rising energy costs. Carriers in the United States typically reference the U.S. Department of Energy fuel price index as an objective resource. When fuel prices exceed an agreed-upon standard, the surcharge kicks in. Surcharges might also come into play when carriers must pay tolls to access roads or rights of ways. Motor carriers often apply surcharges to deliveries in the busy downtown district of cities, where deliveries are likely to take longer.

Finally, *accessorial fees* refer to charges applied to the use of any special equipment or labor employed in support of a shipment. Many shipping and receiving locations are not equipped with loading docks, for instance. In these circumstances, a carrier might offer liftgate service to support the loading and unloading of freight. Some loads might require additional blocking and bracing to secure the load to prevent movement and damage in-transit. When chemicals are shipped via tank truck or rail tankcar, it might be necessary to clean the tank before the next shipment. Some loads also might require additional packaging, hazardous material handling, or high security provisions. In all these instances, the carrier can charge accessorial fees for the additional service provisions.

Summary

In summary, carriers experience accounting costs and opportunity costs in the provision of transportation services. Whereas some costs are fixed, others are variable in nature. Economies of scale found in shipment distance and volume affect carriers costs and, in turn, customer pricing in different modes. Carriers charge fees not only for point-to-point transportation, but also for a variety of reasons, ranging from supplementary service offerings to penalties for delays and the need to cover inflationary economics.

Key takeaways from this chapter include:

- The cost of service is a primary input to determining the prices charged for services.

- Carriers consider both accounting costs and economic costs in managing their operations and serving customers.

- Fixed and variable costs factor into the profitability of carrier operations.

- Carriers are wise to consider the cost to serve customers when pricing services. Activity-based costing (ABC) is a method for determining the cost to serve distinct customers.

- Transportation rates are expressed in different ways, including fixed rates, variable rates, and class rates.

- Carriers might offer several services aside from the basic transport (linehaul) and apply different charges for those services.

Endnotes

1. All carriers employ a minimum charge for shipments, even for small volumes and short-distance moves, to cover the accounting costs and opportunity costs of the move.

2. Chapter 5, "An Overview of Transportation Management," has more on consolidation strategies.

3. "Toot-Toot. Germany Wants More Truckers," *Bloomberg Businessweek* (2 September 2013): 20–21.

4. Truckload Carrier Association, *Dry Van Drivers Survey,* June 1999.

4

THE TRANSPORTATION
SERVICES MARKET

The market for transportation services is characterized by a diverse assortment of providers. They range from the asset-based third-party logistics service provider that is fully responsible for performing the full realm of logistics services, to the non-asset-based broker that maintains relationships with asset-based providers and subcontracts the work. The presence of such a diverse array of service providers in the transportation market asserts that each has a valuable role to fulfill in the marketplace. Furthermore, the introduction of the Internet and other advanced technologies has introduced new forms of providers in the market. This chapter illustrates some of the different forms of service providers found in domestic and international service arenas.

Private Transportation

The first option available for a shipper interested in transporting goods from one place to another is to do it in-house. This is also called *private transportation*. Private transportation employs one's own means to move freight. Technically, it is defined as not-for-hire transportation of a firm's goods, with the firm also owning (or leasing) and operating the transportation equipment for the furtherance of its primary business. This means that, *typically*, private carriers do not provide freight services to the general public (although they are allowed to do so in many nations). Formally speaking, to be considered private, the company cannot be in the business of transportation as a *primary* form of business. Private operations are found in the different modes of transportation.

Private Road Fleets

By far, the most common form of private transportation among shippers is found in road transport. Back in the days of regulation, more companies had private road-based fleets. The reason for this is rather intuitive: When markets offer sufficient options and competitive prices, private operations are less appealing because shippers can buy flexibility at a lower cost, without the distraction from the primary business. However, because markets were tightly controlled during regulation and prices were rarely (if ever) market driven, private fleets were financially worthwhile. In the modern business world, the value of private fleets has diminished somewhat, with so many carriers competing aggressively for business in the marketplace. Nevertheless, some companies continue to maintain fairly substantial private fleets for internal (and sometimes external) use. It is estimated that there are more than 33,000 private fleets of ten vehicles or more in the U.S. alone. Table 4-1 provides a listing of the largest private fleets in the United States, along with the number of vehicles in each (as of 2013).

Table 4-1   The Largest Private Fleets in the United States

Tractor Fleets			Straight Trucks			Overall—Powered Units		
Rank	Company	Fleet Size	Rank	Company	Fleet Size	Rank	Company	Fleet Size
1	PepsiCo	14,293	1	AT&T	63,592	1	AT&T	63,647
2	Coca-Cola	8,200	2	Verizon	58,768	2	PepsiCo	62,066
3	Sysco	7,309	3	PepsiCo	47,773	3	Verizon	58,818
4	Walmart	6,142	4	Comcast	37,087	4	Comcast	37,087
5	US Foods	5,241	5	Waste Management	27,000	5	Waste Management	28,600
6	Halliburton	4,301	6	Time Warner	18,973	6	Time Warner	18,974
7	Nabors Industries	3,920	7	Republic Services	18,149	7	Republic Services	18,343
8	McLane Co.	3,503	8	Century Link	17,924	8	Century Link	17,926
9	Crop Prodn. Services	3,217	9	Cintas Corp.	13,300	9	Coca-Cola	14,100
10	Schlumberger Ltd.	3,062	10	The Service Master Co.	12,784	10	Cintas Corp.	13,400

Source: http://fleetowner.com/%5Bprimary-term%5D/2013-fleetowner-500-1-99.

Companies ranging from small retailers to global corporations often own a fleet of private road vehicles. For example, Walmart has a captive private fleet of more than 6,000 tractors and 55,000 trailers, which together drive more than 750 million miles annually. Similarly, companies such as PepsiCo and Comcast operate substantial private fleets to service their stores, distribution centers, and customer locations. In the recent past, the trend has been to move away from having a private fleet, but there are still some advantages to operating a private road fleet.

Advantages of Private Road Fleets

Arguably, the advantages of owning and operating a private fleet have diminished substantially following the deregulation of the freight business, but some advantages still make it worthwhile for companies to operate their own fleets. This is especially true when the volume of transport required is high. The first advantage relates to potential cost benefits. Rarely do private-fleet operators claim cost savings as the primary reason for running a private fleet, but in some cases, private fleets can provide the shipper with substantial cost benefits over shipping purely through public carriers. This is because such shippers get a credible in-house shipping solution, and potential transporters and trucking companies then need to compete with this in-house option to gain the shipper's business. The shipper is guaranteed the best rates in the open market. In addition, large carriers often return empty from a delivery run (also called *deadheading*), and this cost is baked into the price quote provided to the shipper. Owning a private fleet allows the shipper to pick up shipments from suppliers on the way in, thus providing a cost-effective and environmentally friendly transportation solution for inbound shipping.

Another cost advantage that private fleets provide is that they allow companies to keep their landed costs of goods lower than that of the competition. Walmart, for example, uses its extensive private fleet to collect shipments from suppliers on the way back from the retail store to the distribution center. Because most shippers use *volume of product* in freight lanes as leverage when negotiating freight rates with carriers, Walmart is "pulling" freight volume away from these shippers' freight lanes (by taking over inbound shipping), which lowers the shipper's volume discounts and thereby increases their shipping rates on the remaining shipments. These cost increases then get passed on to other retailers, thereby giving Walmart a substantial cost advantage over the competition.[1]

In addition, in some cases, private fleets are worthwhile simply because of the nature of the cargo or the business. Some shippers have cargo that does not fit neatly into a single category; in such cases, common carriers often end up filing incorrect claims reports. Having a private fleet eliminates this problem. Similarly, when a shipper operates a private fleet, drivers are not merely people who move goods from place to place anymore, but they also act as representatives of the company and, in some cases, even salespeople. Given how much face-to-face contact drivers end up having with downstream channel partners, drivers often become a shipper's best customer service representative. In addition, some shippers feel that highly visible branding on their private fleets rolling down interstate highways provides them with a free billboard, reminding customers of their products and services. In fact, the average truck in an urban area generates an estimated 12,000,000 "viewer impressions" per year.[2]

Some shippers use private fleets because of the comfort they provide in terms of control and guaranteed capacity. (This is especially true in the market of perishable goods.) During peak shipping seasons, it might be hard for a shipper to obtain the necessary capacity in the common carrier market. Controlling a private fleet provides the comfort and convenience of knowing that freight can get to its intended destination in time.[3]

Finally, private fleets afford a level of security that for-hire carriers find difficult to match. Because private fleets are dedicated to serving the company, they are fully aligned with the interests of the company. This includes the safety and security of the loads. Shippers of highly valuable or dangerous cargo often elect to handle transportation through private means. This is true of most companies active in "cash logistics," such as Brinks, Dunbar Armored, and Garda, which specialize in the transportation of money and other valuables.

A middle-ground approach that is gaining much credence is the *dedicated fleet service*. For-hire carriers offer dedicated fleets that look and act as if they were the shipper's private fleet. In some cases, a for-hire carrier acquires the trucking assets and even the labor force once it is employed by the shipper, assuming operations and charging the shipper for the services. Such an arrangement enables the company to combine the benefits of private fleet ownership with the freedom to focus on the company's core (nontransportation) business. Many dedicated fleets maintain the appearances of the shipper, including having the company's logo emblazoned on the trucks and trailers, and even having drivers appear and act as if they are employed directly by the shipper. For example, Exel Logistics provides home and office delivery for different retailers. In the eyes of customers, Exel delivery people appear to be employees of the retailer—they are knowledgeable about the products and can install household appliances on behalf of the retailer. In sum, dedicated services offer an alternative to private fleets and traditional for-hire services to shippers that are large enough to garner the interest of carriers that can provide the service.

Other (Nonroad) Private Fleets

As discussed, private transportation almost exclusively refers to private road fleets because most other forms of transportation are almost exclusively on a for-hire/common carrier basis. Exceptions do exist, however, including the quasi-private ownership of privatized railroad. As such, there are no private railroads in the United States, in the true sense of the term. However, the industry does follow the practice of having private railcars hooked onto and moving by way of common-carrier railroads. In fact, as of 2008, more than 50 percent of the freight-tons shipped on the North American railroads were moved in cars owned by nonrailroad leasing companies and shippers (private railcars).[4] A similar idea is used in water-based transport, where private ownership is typically restricted to barges rather than actual powered vessels. Large companies in the agricultural and energy sectors often maintain their own ports, particularly on rivers and lakes, for dispatching and receiving cargo.

Pipeline transportation is not common among shippers, but this mode is often owned and operated by companies for private use. Technically, the pipelines are considered to be common carriers in the United States, although it is not practical for other firms to use another company's pipeline. Finally, private air carriers are almost exclusively used for transporting people, not freight.

Outsourcing Transportation

The second option available to a shipper who chooses not to engage in self-transportation of goods is to outsource the activity to transportation specialists. In such cases, a shipper might choose to hire the services of a third-party logistics (3PL) firm[5] or to go directly to a carrier (such as a truckload or less-than-truckload carrier).

Contract Carriage (2PLs)

As Chapter 2, "A Survey of Transportation Modes," explained, several types of specialized service providers handle freight movement. *Asset-based carriers* (meaning freight companies that own trucks, tractors, and so on) that engage in the movement of freight and sell their services directly to customers fall into this category. As we have seen, in road-based transport, these are classified as truckload (TL) and less-than-truckload (LTL) carriers. Similarly, in ocean shipping, these are liners and tramp services. Some shippers choose to deal directly with such service providers, who are also called 2PLs (second-party logistics service providers). In many cases, larger shippers deal directly with such 2PLs because of their buying power and large volume discounts. Smaller shippers often must work through brokers and intermediaries because of their smaller volumes and lack of buying power in the market. This creates a market for intermediaries such as 3PLs, 4PLs, freight forwarders, and brokers.

Third-Party Logistics Providers (3PLs)

Basically, a 3PL provider can be defined as an external supplier/vendor that performs all or part of a company's logistics functions. The 3PL side of the freight business is one of the fastest growing aspects of transportation: Between 2010 and 2011 alone, the overall global 3PL marketplace grew from an estimated $541.6 billion to $616.1 billion—a growth of almost 14 percent in one year.[6] More than 86 percent of U.S. Fortune 500 companies use outsourced 3PL services. This growth in the 3PL marketplace is applicable not just in the United States, but all over the world; the Asia-Pacific 3PL marketplace has been growing at an annual rate of 14 percent since 2006.[7] Table 4-2 lists the top ten 3PLs in the world.

Table 4-2   The Largest 3PLs in the World (As of May 2012)[8]

Rank	Company	2011 Gross Revenue (USD in Millions)
1	DHL Supply Chain	32,160
2	Kuehne + Nagel	22,181
3	DB Schenker Logistics	20,704
4	Nippon Express	20,313
5	C.H. Robinson Worldwide	10,336
6	CEVA Logistics	9,602
7	UPS Supply Chain Solutions	8,923
8	Hyundai GLOVIS	8,588
9	DSV	8,170
10	Panalpina	7,358

Three factors are frequently credited with the advent of the so-called "one-stop shop," or the comprehensive, multifunctional 3PL. The first factor is the heightened strategic importance of logistics. Through strategic initiatives such as just-in-time (JIT) inventory systems and efficient consumer response (ECR), which gained popularity in the 1980s and 1990s, companies sought advanced logistics expertise and capabilities that were not found in shipper organizations. Out of this need, traditional logistics service providers, such as warehouse operators, transportation carriers, and brokers, expanded their portfolio of services to the comprehensive, one-stop shop variety.

The second factor that supported this service evolution is massive deregulation in the logistics industry, particularly deregulation of the transportation sector in the United States. The Motor Carrier Regulatory Reform and Modernization Act of 1980 (MCA-80) was instrumental in allowing warehouse companies to more freely engage in transportation operations and, conversely, for transportation carriers to offer broader, integrated forms of logistics service. Furthermore, MCA-80 allowed creativity and innovation to enter the logistics industry following nearly a century of increasingly restrictive regulation. Deregulation is credited with inciting similar market freedoms in settings outside the United States as well, and reinforcing development and adoption of outsourced logistics services in these various settings.

A final yet key factor that has contributed to the growth of 3PLs is increased competition and globalization in the twenty-first century. Booming world trade has fostered longer and more complex supply chains which, in turn, have created a pressing need for better logistics in all corners of the world. For example, cities such as Coimbatore in India and Chittagong in Bangladesh house several small workshops and factories where several of the world's leading brands source garments. However, in many of these places, the logistics infrastructure can be quite primitive. The major retailers and manufacturers outsourcing to these factories are using their relationships with large 3PLs to control how their overseas vendors perform their logistics. Similarly, growing prosperity in rapidly developing parts of the world, such as Asia, Africa, and Eastern Europe, is building thriving consumer markets. Large manufacturers in Europe, the United States, and Japan want to ramp up their activities in these regions to position themselves for this new wave of consumerism. To accomplish this, they often turn to specialist 3PLs to gain access to those markets. As such, 3PLs can be thought to fall under two broad categories: asset-based and non-asset-based:

- **Asset-based 3PLs**—Asset-based 3PLs own and operate the elements of the supply chain being used to service logistics needs. These providers generally own warehouses, support equipment, and, most important, over-the-road equipment such as transport trucks. They market their services directly to shippers and seek to utilize their own elements in the supply chain, such as their own transportation vehicles and warehouses. They commit to serve the customer and, by virtue of performing the services themselves, are directly accountable to customers for the service they provide, including responsibilities for over, short, and damaged (OS&D) claims.

- **Non-asset-based 3PLs**—Unlike asset-based 3PLs, non-asset-based 3PLs do not own the assets necessary to manage transportation. Thus, instead of offering use of their tangible assets (such as trucks, trailers, and warehouses), non-asset-based 3PLs offer their skills and expertise in finding the best providers for all these activities, negotiating contracts with these vendors, and implementing the solution. In essence, non-asset-based 3PLs are more like transportation consultants than actual freight movers. Distinction must be made, however, between brokers and non-asset-based 3PLs because 3PLs often assume primary liability for cargo claims and indemnity arising from losses and damages. It is critical that such understanding be achieved in all contractual service arrangements with non-asset-based service providers who will ultimately subcontract the work to asset-based carriers.

- **Choosing between asset-based and non-asset-based 3PLs**—Considerable debate arises on the advantages and disadvantages of asset-based versus non-asset-based 3PL service providers. Each shipper has different needs, so no single approach applies to all shippers. Most shippers typically look for a combination of one or more of the following factors in making their choice on asset-based versus non-asset-based 3PL partners:

  - **Cost**—Eventually, one of the key reasons that shippers choose to engage a 3PL in the first place is to be able to serve their customers (current and future) as effectively as possible at the lowest cost. Therefore, potential cost savings provided by the 3PL are an important aspect of the overall decision framework for most shippers. Given that asset-based 3PLs own some or all the physical assets necessary to the transportation requirement, they can sometimes offer a lower cost on warehousing and transportation: They set their own pricing and are not paying an outside party, such as a motor/rail carrier or warehouse operator. This is especially true if the non-asset 3PL does not have the buying power when it comes to buying the provisions of transportation and warehousing capacity.

- **Flexibility**—The advantage sometimes enjoyed by asset-based 3PLs can be potentially negated because of the second dimension: flexibility. Asset-based 3PLs use their own assets, such as trucks, trailers, and warehouses, to move the shipper's freight, so they are sometimes constrained in terms of the range of solutions they can provide. Non-asset-based 3PLs, however, can work with shippers to identify the best-in-class solutions to meet the shipper's needs. Of course, asset-based 3PLs that are large enough often have enough options within their own system to be able to accommodate all the flexibility that a shipper needs, so that point then becomes irrelevant. From an economics perspective, however, developing expansive capabilities and capacities comes at a cost. Asset-based providers have bills to pay whether the diverse assets they own are fully or partially utilized. Hence, asset-based providers must charge prices that ensure profitability whether assets are employed fully or not, meaning that flexibility comes at a price.

- **Continuity**—Asset-based 3PLs are often considered more stable than non-asset-based 3PLs. There is some truth to this belief: The current surety bond for beginning business as a 3PL is merely $75,000, and often some 3PLs are little more than a small office with a skeletal staff, a phone, and an Internet connection. This lack of an entry barrier makes it easy for fly-by-night non-asset 3PLs to easily close up shop, leaving bills unpaid and customers hanging. Thus, the pressure on the shipper is sometimes higher when hiring a non-asset-based 3PL (especially a new one).

Lead Logistics Providers/Integrators (4PLs)[9]

The explosive growth in the 3PL marketplace across diverse operating geographies has given rise to firms that often have relations with multiple 3PLs at a time. For example, as of 2013, Procter & Gamble and Walmart maintain relationships with more than 50 3PLs each. Often these 3PLs manage different subsegments within a transportation network. This problem is more prevalent in some industries. For example, the grocery retailing business is known for supply networks that are highly regionalized, to more quickly source a wide array of local specialties just-in-time. In such a case, grocery firms are highly decentralized and operate on different enterprise systems with their 3PLs in every market they serve.

This has given rise to a market for a new type of freight intermediary: a *fourth-party logistics service provider (4PL)*, also sometimes called an *integrator*. A 4PL is a logistics management service that does not touch or run logistics services itself. Instead, it manages multiple 3PLs and freight providers on behalf of the vendor. Given how new the term is, some ambiguity surrounds the meaning of the term itself, and this will probably be the case for some time. Therefore, in this text, we prefer to stick with the following definition of 4PLs: logistics intermediaries that are charged with coordinating the operations of a collection of logistics service providers. Obviously, a 4PL is a highly specialized and complex transportation intermediary, not one that many businesses (especially small and medium ones) immediately require.

A distinct form of service provider is the *lead logistics provider (LLP)*. Unlike the 4PL, the LLP can be an asset-based provider that actively manages the freight. The LLP serves as the primary owner of the relationship with the shipper, yet it hires additional carriers and logistics service providers to conduct operations in support of the shipper. Such arrangements are quite common in the inbound side of automotive manufacturing. Automotive manufacturers often hire a single large 3PL to serve as the LLP. This LLP coordinates the inbound logistics system, arranging the flow of parts and components to feed manufacturing. This 3PL performs some of the operations itself and hires out for the rest, assuming responsibility for the performance of the entire inbound system. Similar arrangements are found around in the world in fashion, electronics, pharmaceutical, and chemical manufacturing and distribution.

Freight Forwarders

To understand who *freight forwarders*[10] are, we need to explore both the technical definition and then services provided. The U.S. Federal Motor Carrier Safety Administration (FMCSA) defines a freight forwarder as follows: a person or entity holding itself out to the general public (other than as a pipeline, rail motor, or water carrier) to provide transportation of property for compensation in the ordinary course of its business such that it does the following:

1. Assembles and consolidates (or provides for assembling and consolidating) shipments, and performs (or provides for) break bulk and distribution of the shipments

2. Assumes responsibility for the transportation from the place of receipt to the place of destination

3. Uses any part of the transportation of carrier subject to jurisdiction under this subtitle

In essence, freight forwarders are logistics service providers that serve as intermediaries between the shipper and the carrier, and that can carry out several different activities for shippers. First of all, most freight forwarders have some method of storing the client's cargo at their own warehouses or affiliated warehouses. Along with storing goods, freight forwarders sometimes can sort, consolidate, and break bulk; negotiate freight rates with shipping lines to cover the interests of their clients; book the cargo with the shipping line; and prepare shipping documents, such as bills of lading.

A key point to note (and this is what sets freight forwarders apart from other intermediaries, such as freight brokers) is that freight forwarders are authorized to issue their own bills of lading. In addition, in the United States, freight forwarders have primary liability to the shipper for cargo loss and damage, just as any receiving carrier has. This means that freight forwarders have full responsibility (and liability) for the shipment from the time of initial receipt to final delivery. In addition, their special status as a freight intermediary means that they are responsible for payments to the carriers. Thus, in effect, freight forwarders are the true intermediaries of the freight business because they act as shippers vis-à-vis the carrier, and carriers vis-à-vis the shippers. This is why freight forwarders must carry cargo and liability insurance policies equal to at least the minimum coverage amounts required for carriers.

In the United States, freight forwarders need to be registered with the FMCSA. Some freight forwarders might also be affiliated with organizations such as the International Air Transport Association (IATA). Generally, freight forwarders have large-scale relationships with both ocean freight and air freight carriers (although some freight forwarders specialize in one of these categories). Because of their substantial buying power, they usually can offer considerable cost savings to their customers. They also can assist in choosing the right carrier for the shipment and can handle the support documentation. Thus, from a shipper's perspective, little difference exists between freight forwarders and shipping lines themselves. Indeed, at times, shippers view freight forwarders as an alternative shipping line.[11]

Difference Between 3PLs and Freight Forwarders

Given our earlier discussion of 3PLs and freight forwarders, the next logical question to ask is whether there is a difference between the two. Traditionally, the distinction between these two service providers used to be clear: Freight forwarders moved freight from airport to airport or port to port by using shipping and airlines, and 3PLs took care of warehousing and distribution. Over time, however, the distinction between these two has been blurring. Traditional freight forwarders have begun to offer a broader array of services and have rebranded themselves as 3PLs. Therein lies the key difference: Typically, 3PLs offer a broader array of services than freight forwarders.

The basic difference between the two is that freight forwarders are specialists that focus on the logistics of transportation, and they provide *some* extra services (such as warehousing, inventory management, and documentation support). 3PLs, on the other hand, are usually generalists who help move the goods, but their focus is not merely on goods movement; they add several extra services, such as storing goods between transport steps, processing shipments, taking items from transport vehicles and packaging it with an invoice per client drop off, and providing visibility into the channel, among a host of other activities. In addition, freight forwarders typically focus on smaller linkages in the supply chain instead of trying to optimize the entire supply chain. Finally, whereas 3PLs can be either asset-based or non-asset-based, freight forwarders have traditionally been only non-asset-based (although this is changing). So in essence, the difference is that freight forwarders have a narrower focus on the *transportation* aspect of the supply chain, whereas 3PLs have a more *broad focus*.

Brokers

A fair amount of confusion arises when considering freight brokers and freight forwarders. The term *freight broker* means a person or entity other than a motor carrier, or an employee or agent of a motor carrier, who arranges for the truck transportation of cargo belonging to others, utilizing for-hire carriers to provide the actual truck transportation. However, the broker does not assume responsibility for the cargo and usually does not take possession of the cargo.

Consider this easy way to understand that definition: If an entity is allowed to issue a bill of lading and take over the liability for the goods being transported, that entity is something *other* than a broker. Stated differently, brokers are *not* allowed to issue their own bills of lading with their names in the carrier field, unlike freight forwarders. (If a shipment is arranged with a broker instead of with a freight forwarder, the actual carrier's name is mentioned on the bill of lading.) In addition, brokers are not responsible for any freight claims and can merely help in forwarding such claims (if they occur) to the carrier for handling or compensation. Finally, brokers do not provide any insurance coverage for the cargo. Thus, a freight broker is merely a conduit for connecting the shipper and carrier.

Because freight brokers take much lower amounts of liability and responsibility for freight than freight forwarders, the entry barriers to this business are also typically lower. A freight broker must be registered with the Federal Highway Administration and must post a surety bond or trust fund in the amount of $75,000. In addition, brokers are required to keep a record of each freight transaction for a minimum period of three years. These records must show the name and address of the consignor, the motor carrier, the bill of lading or freight bill number, the amount of the broker's compensation, a description of any non-brokerage services performed in connection with each shipment or other activity, the amount of any freight charges collected by the broker, and the date the payment was made to the carrier.

Summary

In sum, service providers in the domain of transportation management assume a wide variety of forms and names. These terms can sometimes be confusing. It is critical that supply chain managers achieve clarity in the service arrangements that they devise with different types of providers. As the previous discussion indicated, service providers assume different responsibilities for liability, depending on form. Understanding and responsibly assuming risks is among the greatest concerns in transportation management. Hence, it is essential that these differences are properly acknowledged in transportation transactions.

Key takeaways from this chapter include:

- The most common form of private transportation is through private road fleets.
- Contract carriage is common among large shippers because they have regular recurring needs and considerable shipping volumes.
- Factors leading to the rise of 3PLs include an increased focus on logistics, deregulation, and globalization.
- 3PLs can be divided broadly into asset-based and non-asset-based providers.
- 3PLs are typically selected based on cost, flexibility, and continuity.
- Integrators/4PLs are a recent and growing intermediary.
- Freight forwarders are distinct from brokers and 3PLs. All three service provider forms fulfill unique needs for shippers in supply chain operations.

Endnotes

1. Penn State's Smeal College of Business, "Wal-Mart's New Transportation Strategy." *Business Casual.* June 7, 2010. http://blogs.smeal.psu.edu/businesscasual/2010/06/07/walmarts-new-transportation-strategy/.

2. Statistic from Mobile Billboard Advertising.

3. John Edwards, "Private Fleets: Your Own Private Ride." *Inbound Logistics.* September 2006. Available at www.inboundlogistics.com/cms/article/private-fleets-your-own-private-ride/.

4. Thomas M. Corsi, Ken Casavant, and Tim A. Graciano, "A Preliminary Investigation of Private Railcars in North America." *Journal of the Transportation Research Forum.* 2012;51(1):53-70. Available at www.trforum.org/journal/downloads/2012v51n1_04_Railcars.pdf.

5. Depending on prevailing business practices, some firms also choose to use the terms *logistics outsourcing* or *contract logistics* instead of *3PL*.

6. Joe Lynch, "Capgemini's 2012 3PL Study Provides Great Insights." *The Logistics of Logistics.* November 6, 2013. Available at www.thelogisticsoflogistics.com/2012/11/capgeminis-2012-3pl-study-provides-great-insights/.

7. Josh Bond, "3PL customers report identifies service trends, 3PL market segment sizes and growth rates." *Modern Materials Handling.* July 22, 2013. Available at www.mmh.com/article/3pl_customers_report_identifies_service_trends_3pl_market_segment_sizes_and.

8. Patrick Burnson, "The Top 50 Global and the Top 30 Domestic 3PL Providers." *Supply Chain 24/7.* March 17, 2013. Available at www.supplychain247.com/article/the_top_50_global_top_30_domestic_third_party_logistics_providers.

9. Accenture originally coined the term *4PL* and registered it as a trademark in 1996.

10. In some countries, freight forwarders are known as clearing and forwarding (C&F) agents.

11. A directory of freight forwarders is available at www.forwarders.com.

Section 2

Transportation for Managers

5

AN OVERVIEW OF TRANSPORTATION MANAGEMENT

The first section of the book provided a foundation for understanding the domain in which transportation decisions are made and also previewed some of the options available to shippers. We reviewed the various modes of transportation, the economics of transportation operations, and the different forms of service. In this chapter, we synthesize the decisions that shippers face in managing the transportation function in their businesses. The decisions range from the strategic decisions associated with network design, to the tactical considerations of how to best load a transportation vehicle. Network design decisions crop up only infrequently when the dynamics of the business change, as with the pursuit of new markets or when shipping volumes change among existing markets. On the other hand, load preparation and routing decisions occur with each shipment. This chapter reviews this range of decisions, with an emphasis on how transportation interfaces with activities and decisions made in the domains of logistics and supply chain management (SCM). Achieving integration among the various activities is essential for a high-performing company and supply chain. A decision-making framework is introduced to organize the flow of decisions from most strategic to tactical in nature.

Transportation Management Decision Making

In Chapter 1, "Transportation in Business and the Economy," we established the economic and strategic significance of transportation toward the competitiveness and success of a company. Transportation is ordinarily the largest single cost across all the logistics activities and thus rates among the most important service dimensions for the company. Excellent transportation service can portray the company as a reliable supplier. Poor service suggests that the company is one to avoid.

In light of the vast array of decisions associated with transportation strategy and operations, it is important to organize the decisions in some manner. Figure 5-1 illustrates a decision flow that first concerns strategic decisions, such as network and lane design, and then becomes more operational in nature. As the decision scope suggests, the more strategic decisions tend to be macro, or large in scope, and affect many other decisions. The micro decisions are not less significant, but they focus more on specific shipments. The next five sections of this chapter review these decision areas in detail.

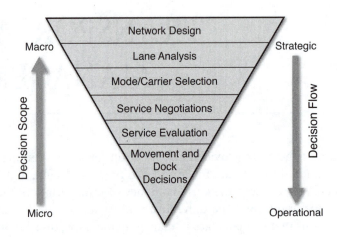

Adapted from: Stank, Theodore P. and Thomas J. Goldsby (2000), "A Framework for Transportation Decision-Making in an Integrated Supply Chain," *Supply Chain Management: An International Journal,* Vol. 5, No. 2, pp. 71-77.

Figure 5-1   Integrated transportation decision-making framework.

Network Design

The highest level of decision making and analysis in transportation management involves determining the locations for facilities in the *network design.* The network consists of the shipper's array of facilities, whether manufacturing, distribution, or retail in nature. The facilities serve as the company's physical presence in the supply chain, linking the company to suppliers (upstream) and customers (downstream). Many companies operate multitier distribution channels as well, with the firm essentially serving as its own sources of supply or points of use/consumption. Transportation must be coordinated within these internal supply chains, too. The essence of decision making at this high, strategic level is to achieve the best possible connections that meet the service needs of the focal company on the inbound side of the business and allow the company to successively serve customers downstream—and at the lowest possible cost.

Some companies centralize their inventory, meaning that they hold inventory in one or a few locations to serve a large market. This makes determining where to position inventory easy. However, it does mean that customers in far-flung regions might be disappointed if they have to wait for deliveries. Imagine, for instance, that your company wants to serve customers throughout South America. You elect to hold all your inventory in one location, Sao Paulo, Brazil. Sao Paulo is the largest city on the continent and holds a central latitude for the South American market. Customers in Sao Paulo and nearby Rio de Janeiro would agree that the choice is a good one. However, customers in distant Bogotá, Colombia, or Santiago, Chile, might think otherwise: They would face long order lead times (the elapsed time from order placement to delivery). For this reason, most businesses elect to stock inventory in several disparate locations to serve a large market such as South America.

The general premise of balancing service and cost has an extensive history. Many date the scientific approach to network design to the seminal problem posed by Euler (1735), known as "Seven Bridges of Konigsberg." Even though Euler could not solve the problem of walking through the town (present-day Kaliningrad, Russia) by crossing the seven bridges of River Pregel/Pregola only once, he laid the foundation of using graph theory to analyze transportation problems. During the twentieth century, *graph theory* gave way to network analysis because of its systems approach.

Transportation systems are considered spatial networks because of the physical limitation of their designs. A transportation system often consists of a *node,* which represents a location. The *flow,* or the amount of traffic of a node, is often depicted by an arc that links the nodes.

Typology of Transport Networks

This section reviews different forms of transport networks:

- **Hub-and-spoke networks**—Transportation networks that rely on a few large nodes to direct and redistribute traffic to smaller nodes are often referred to as hub-and-spoke networks or a star network. At the very extreme, an N node network can have N-1 links to connect every node with a hub. Airlines, railroad, and freight companies use the hub-and-spoke model to gain economies of scale in their operations. The hub-and-spoke model allows for new nodes (spokes) to be created easily and connected to the main hubs. The hubs aggregate the traffic of freight or passengers from smaller nodes and tend to have high volume and high frequency of traffic between the hubs. One of the biggest disadvantages is the creation of a bottleneck, a single point of failure that can cripple the network. The time required to move freight/passengers to their destination is higher than in a point-to-point network. Airbus is taking advantage of the hub-and-spoke model to fly passengers across the world by building the A380 and A350 models of planes.

■ **Point-to-point networks**—With point-to-point networks or mesh networks, each individual node is connected to every other individual node. At a maximum, the number of links in an N network node is N(N-1)/2. In a transportation network, only nodes that are efficient in terms of cost and that carry substantial traffic connect to each other. For a long time, Southwest Airlines connected routes on a point-to-point basis. The advantages of this form of network are the elimination of a big central hub and reduced travel time. A big disadvantage of this methodology is the decreased frequency of trips and the inability of very small nodes to connect to a larger network. Boeing is relying on its 787 models of airplane to connect point-to-point nodes. Figure 5-2 illustrates the differences in hub-and-spoke and point-to point systems.

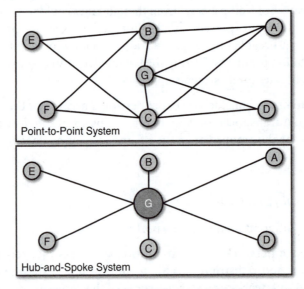

Figure 5-2   Hub-and-spoke system versus point-to-point system.

■ **Distributed networks**—The distributed network consists of nodes that are connected based on prevailing demand and supply equilibrium. No routes or schedules are fixed. Freight and passengers are taken from one node to another based on the availability of demand and the ability of the transport provider to generate revenues in excess of the costs. Sea and air charters are excellent examples of distributed networks. The big disadvantage of a distributed network is the inability to predict accurate demand in advance and, hence, plan for upgrading infrastructure at the nodes.

Other ways of classifying networks include methods focused on the flow of traffic:

- **Centripetal networks**—These networks occur when the center of gravity of different flows aggregates to a point. For example, the city of Washington, D.C., has its center of gravity located near The Mall, where tourists gather in large numbers and where the seat of the government attracts employees and businesses. This tends to give the city a radial pattern of flow, with smaller nodes feeding into the large center of gravity.

- **Centrifugal networks**—The flows seen on a centrifugal network resemble a grid pattern, with no specific node dominating the landscape. New York is an example of a centrifugal network—no specific link dominates the traffic, and the grid pattern prevents a build-up of traffic at any intersection. Figure 5-3 illustrates the differences in centripetal and centrifugal networks.

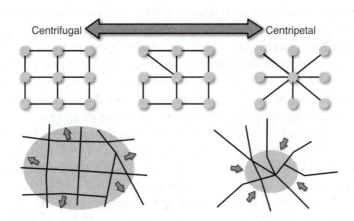

Figure 5-3   Centrifugal network versus centripetal network.

The various models currently being used to solve transportation are described next.

Optimization

When solving transportation problems specifically to allocate resources between a set of origins and a set of destinations, optimization is commonly the preferred methodology. The objective of the model is often the cost incurred in transporting the resources from each pair of origin-destination. The constraints are typically defined in terms of the capacities at the origin, the destination, intermediary points, and the links. The main advantage of using optimization techniques is the ability to evaluate all possible routings before making the optimal choice. The disadvantage is that the model needs to be as simple as possible to avoid tying up resources of time and cost to solve the model. Also, not all constraints can be modeled accurately.

Heuristics and Simulations

Transportation problems are often solved using "rules of thumb" or *heuristics*. The most likely outcome is often modeled as a Monte Carlo simulation to set up a base scenario. The data is then manipulated to look at conditions of high or low traffic flows and the effect on time, money, and congestion. The what-if scenarios generated by the simulation help policy makers and planners anticipate any shortcomings in the network and take corrective action. A big disadvantage of the methodology is the danger of "garbage in, garbage out": The integrity of the solution is only as good as the efficacy of the rules used in generating the solution. Bad data also tend to skew results in the wrong direction.

Regardless of the method employed, when optimization is achieved and implemented, it becomes challenged immediately. Network analysis often occurs in a single moment of time (although the analysis might require several months to complete), and a design that was once optimal might have shortcomings when it is implemented, given the dynamics of business. Network designers often make an inherent assumption that what has happened in the past will continue into the future, or they adapt their forecasts to reflect anticipated volumes and shipping points. As noted, many companies employ heuristics and simulations as supplemental means of assessing alternative network designs. Yet without question, the decisions of 1) how many facilities (nodes) to design into the logistics network and 2) where to locate the facilities are among the most substantial and critical a company can make. In real estate, it has long been believed that the three most important components of success in that business are location, location, location. The same can be said of SCM, for it not only has service and cost implications for the business, but it also conveys an important message to customers. Critical customers often require stockkeeping locations near them, and the company's own marketing and sales organization backs up this assertion. Yet with each location come fixed costs and the costs of operations.

With modern complexities, however, come solutions. Third-party logistics companies (3PLs) offer a flexible solution to the network problem. Today's 3PLs are often willing to locate wherever a shipper requires for as long as is necessary. In essence, leveraging the facilities of an outsourced party converts networks into *flexworks*. Shippers can more freely enter and exit markets and use more or less space, as needed. These arrangements also convert the company's fixed costs of facilities into semifixed and variable costs.

The semifixed component is associated with the guarantees that a 3PL might require to assume the risk of fixed costs shifting from shipper to service provider. The 3PL would then likely charge for the volume of freight serviced and the activities performed, both on a variable basis.

Lane Analysis

After determining a network, attention turns to the flows among the facilities, including flows from suppliers feeding the network, any intranetwork moves, and flows outbound to customers. Here the network design is put to the test. Again, the premise is to provide service that customers demand at the lowest possible cost.

One way to lower costs is to utilize the vehicle capacity optimally to support the flows. Sending half-full (or less) vehicles into the market for deliveries is not an optimal use of transportation equipment. Similarly, carpooling or ride sharing is suggested for people commuting from nearby homes to the same workplaces, to reduce the wastes of redundant travel; it saves money and reduces congestion. The same idea applies to freight transportation as well. *Freight consolidation* is the practice of combining shipments for improved transportation utilization. Consolidation comes in three forms:

- **Vehicle consolidation**—Consolidating multiple customers' orders with stopoffs along the way for outbound deliveries, or consolidating multiple suppliers' shipments on the inbound side

- **Temporal consolidation**—Advancing or delaying the shipment of an order to allow for consolidation on either the inbound or outbound side of the business

- **Inbound/outbound consolidation**—Coordinating shipment receipt and outbound delivery so that after an inbound shipment is unloaded, an outbound load can be sent on the same vehicle

Vehicle consolidation might involve converting multiple less-than-truckload (LTL) shipments into a single truckload shipment. The truckload carrier will charge its typical rate for the move from the origin to the furthest destination. It will then charge for stopoffs and any out-of-route distance it must cover to accommodate the stops at the intermediate locations. The question is how the charge for the truckload with stopoffs will compare to the sum of multiple LTL shipments. The timing and sequencing of deliveries must also be considered so that each shipment meets its promised delivery.

Temporal consolidation can present opportunities when orders destined for a single customer or multiple customers in a small region are placed at different times, with distinct delivery dates/times. This typically results in distinct shipments for each different order. However, it might be possible to combine the orders into a single shipment, even if it means advancing or delaying one or more orders to make this consolidation possible.

Clearly, the customer should be informed if the carrier entertains a change in the delivery time to accommodate the consolidation. Furthermore, the receiver should approve the prospective change. Including the customer in the cost savings is one way to encourage the customer to consider the joint shipment.

The challenge with these first two forms of consolidation is the difficulty associated with scheduling multiple stops and ensuring that all customer commitments can be met. Imagine, however, if there is a significant delay at the first stop in a consolidated move. This delay is likely to put the remaining deliveries in peril of missing their delivery appointments. Therefore, it is important not only to schedule the deliveries with precision, but also to reinforce discipline around the deliveries at the various stops. Unfortunately, the timely turnaround of transportation assets at delivery points is often beyond the carrier's direct control because the carrier is often at the mercy of the receiver to unload the freight, complete the delivery, and send the driver to the next stop. With this in mind, shippers must carefully consider the customers at the intermediate delivery points that can jeopardize the deliveries to successive locations. Carriers typically learn quickly where they can count on quick turns and where they cannot. This suggests, however, that new delivery locations should probably not factor into consolidated shipments until their reliability is ensured.

Finally, inbound/outbound consolidation is most often found among truckload, rail, and maritime (water) carriers. These carriers specialize in making point-to-point deliveries. The significance of point-to-point delivery is that after a shipment is delivered to a customer, the carrier must often travel to another location to collect freight from another customer. The greater the distance a vehicle must travel to claim the next load, the more costs the carrier incurs. If a customer can send freight from the same location that received a load from the carrier, the carrier will be inclined to offer a discount for providing revenues both coming to and going from the facility. The discounts might apply to the inbound or outbound load, or perhaps both loads. Truckload carriers, in particular, are eager to offer discounts for shippers that offer inbound and outbound loads in concert because they are the carriers most likely to incur empty deadhead (nonrevenue) distance to the next shipper location. To achieve such consolidations and discounts, the shipper must manage both the inbound and outbound flows. If a supplier arranges the company's inbound deliveries, coordination will be lacking—and so might incentives because different parties are likely paying for the inbound and outbound freight.

Note that these three forms of consolidation are not mutually exclusive. For instance, it is possible to design a series of inbound moves that employ vehicle and temporal consolidation at the same time, with inbound/outbound consolidation occurring on the outbound trip.

To illustrate these forms of consolidation, examine Table 5-1. It lists several inbound shipments to a facility in Columbus, Ohio, and an equal number of outbound shipments from this same facility. The weight, promised delivery date, and anticipated form of service are found with each shipment. LTL refers to a shipment that weighs less than 8,000 pounds (in this example) and is expected to ship via LTL carriers. Shipments of 8,000 pounds or more appear to be shipped via truckload (TL) carriers. On the outbound side, some loads are also designated as 1-D, meaning next-day service, and 2-D, meaning second-day delivery, for time-critical deliveries. An understanding of the geography of the United States is helpful in completing the exercise, just as it is essential to understand the lanes over which any inbound shipment or outbound delivery operates. With this understanding, one can explore possible ways to combine shipments for consolidated volumes.

Table 5-1 Find the Opportunities for Consolidation

Origin City	Weight	Delivery	Service	Destination City	Weight (lbs)	Promised	Service
Memphis, TN	14,000	3/27	TL	San Antonio, TX	130	3/28	1-D
Tampa, FL	7,500	3/27	LTL	Baltimore, MD	110	3/29	1-D
Amarillo, TX	4,800	3/28	LTL	Birmingham, AL	6,000	3/29	LTL
Detroit, MI	3,200	3/28	LTL	Charleston, WV	500	3/29	LTL
Shreveport, LA	12,500	3/28	TL	Cincinnati, OH	9,000	3/29	TL
Buffalo, NY	26,000	3/29	TL	Jackson, MS	2,400	3/29	LTL
Chicago, IL	19,000	3/29	TL	Jacksonville, FL	70	3/29	1-D
Pittsburgh, PA	18,000	3/29	TL	Knoxville, TN	5,200	3/29	LTL
St. Louis, MO	16,000	3/29	TL	Charleston, WV	9,500	3/30	TL
Indianapolis, IN	6,850	3/29	LTL	Dayton, OH	110	3/30	1-D
Kansas City, MO	6,000	3/30	LTL	Evansville, IN	8,000	3/30	TL
Nashville, TN	4,500	3/30	LTL	Miami, FL	85	3/30	2-D
St. Louis, MO	14,000	3/30	TL	Newark, NJ	10,000	3/30	TL
Indianapolis, IN	10,800	3/31	TL	Philadelphia, PA	125	3/30	2-D
Shreveport, LA	3,500	3/31	LTL	Charlotte, NC	8,600	3/31	TL

If we assume that a truck has capacity for up to 40,000 pounds of freight, we can identify several prospects for consolidation. Figure 5-4 illustrates some of the possibilities. Vehicle consolidation is possible with the inbound shipments from Kansas City, Missouri, and St. Louis, Missouri, because both are scheduled for delivery in Columbus on March 30. Note that the St. Louis load is already designated as a truckload shipment, given its weight of 14,000 pounds. However, the Kansas City shipment, at 6,000 pounds, is expected to travel by way of an LTL carrier. Instead, we might consider arranging with a truckload carrier to collect the smaller shipment in Kansas City and to travel east toward Columbus. When passing through St. Louis, the carrier could collect the 14,000-pound load and then continue east to Columbus. The carrier would charge for the pickup at the intermediate location (St. Louis), along with any out-of-route miles, yet the cost of this combined shipment would likely be less than the combined cost of the two independent shipments (the LTL shipment from Kansas City and the TL shipment from St. Louis). In addition, the shipment from Kansas City could occur faster via the truckload carrier because it would avoid the rehandling (loading/unloading/reloading) of freight found in conventional LTL environments. This rehandling can also cause damage because the risk of damage increases with each touch placed on freight.

An opportunity for temporal consolidation arises with the outbound shipments to Philadelphia, Pennsylvania, and Newark, New Jersey. The Newark shipment is already designated for truckload delivery with its 10,000-pound load. Yet ample space remains to add the 125-pound shipment to Philadelphia. This shipment was not anticipated to ship via 2-D transit, but Philadelphia can be reached in 1-D from Columbus via standard truckload or LTL means. Companies often employ premium same-day, 1-D, and 2-D services for small-volume shipments only to learn that standard service would serve the purpose at a much lower cost. Such is the case here: The Philadelphia shipment could "ride along" with the load to Newark. As in the earlier vehicle consolidation example involving the inbound loads from Kansas City and St. Louis, a single truckload truck would be used for the Philadelphia and Newark shipments, with the driver stopping in Philadelphia to drop off the small shipment. Again, a stopoff fee and out-of-route distance fees would likely apply, so these costs must be considered in the comparative analysis. Also, this delivery might qualify as a temporal consolidation if it altered the original scheduled delivery times for either shipment. Both shipments have delivery dates of March 30. However, to ensure that the larger shipment arrives in Newark by March 30, it might be necessary to move the Philadelphia delivery date up to March 29, particularly if the Newark customer seeks delivery early on March 30. The adjustment of the delivery date would qualify this combination as a temporal consolidation.

Inbound/outbound consolidation can occur if a truckload shipment bound for Columbus can be turned around quickly upon arrival for an outbound shipment that sends the truck and driver back in the direction of the origin of the inbound shipment. Again, the premise in truckload is to return the truck and driver back to the home terminal. Carriers reward shippers that can provide revenue on these backhaul (return) trips. Such appears to be the case with the March 27 inbound shipment from Memphis, Tennessee. It should arrive in time for the truck to be unloaded and then loaded with the outbound freight destined for Cincinnati, Ohio, on March 29. Each shipment is scheduled to move truck-load carriers. If the customer were to schedule both shipments via the *same* carrier, the carrier might reward the shipper with a discount on the inbound and outbound loads, particularly if the customer can make a commitment to finding these opportunities on a regular basis. Note that the outbound shipment does not return the truck and driver all the way back to Memphis, but goes only partway, to Cincinnati. Yet even this short trip provides the carrier with a convenient source of revenue. Furthermore, it gives the carrier an opportunity to find freight in the Cincinnati area destined for Memphis that would provide yet another revenue trip on the return. In light of the uncertainty that carriers often face on backhauls, carriers welcome convenient opportunities for earning revenue on these trips.

Figure 5-4 illustrates these three different forms of consolidation. Each form of consolidation is designated by a different arrow in the figure. Although not a consolidation, another opportunity for cost saving is shown with the shipment from Columbus to Jacksonville, Florida. Here, a 70-pound shipment is scheduled for 1-D delivery. Yet if the inventory is available on the current day (March 27), it can reach Jacksonville safely by the delivery date (March 29) using standard delivery. This qualifies as a class shift, changing the designation of the shipment from a premium service arrangement to a standard service arrangement. It is not uncommon for premium time-definite deliveries to cost 70 percent more than standard service, thus encouraging companies to find these opportunities to convert premium freight to reliable standard services.

As a final observation of the freight consolidation example, it is often possible to consolidate more than two shipments. Careful review of the shipping data can reveal opportunities for several more consolidations, including prospects for incorporating many different loads into a unified inbound or outbound shipment.[1] It is essential that freight consolidations factor in 1) the timing of the loads, to ensure that the integrity of delivery schedules remains intact, 2) the ability to allow different freights to share capacity, and 3) the observation of weight and capacity limits. Hence, freight consolidation is often conducted with the aid of information technologies, such as a *transportation management system (TMS)* and/or load-planning software, by seasoned transportation professionals who can make accurate calls on the viability of consolidation in light of the three factors.

Consolidation is among the foremost considerations in lane analysis to seek the desired service levels at the lowest possible costs. Routes can change, however, with changes in supply and demand locations and volumes shipped. As an example, Toyota devises regular routes for its inbound logistics service. The volumes and frequency are based on the production volumes at Toyota plants. When production volumes increase, the volume and/or frequency of supply must increase to support the rising production. Inversely, when production decreases, volumes and/or frequency of inbound freight reduces in kind. As a standard practice, the company and its third-party logistics providers review the inbound routes and make adjustments about 16 times each year (about once every three weeks). These adjustments can include the volumes collected at each supplier site, the frequency of pickup, or the assignment of suppliers to different routes. Good lane analysis allows a company to identify opportunities for improved service and cost reduction.

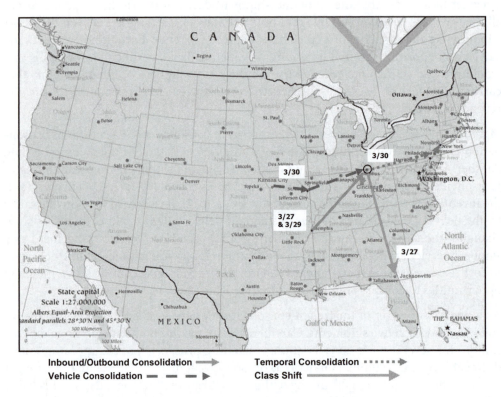

Figure 5-4   Examples of freight consolidation.

Mode and Carrier Selection

In the discussion of network design and lane analysis, implications for mode and carrier selection started to appear. The mode and, ultimately, the carrier the company commits to determine how it will accommodate the shipping need. Figure 5-5 lists several criteria a company considers when choosing a mode or carrier. The level of service in terms of expected transit time (speed) and consistency (reliability) factor significantly into the decision, along with the associated cost. Increasingly, companies are taking greater interest in factors such as security and qualitative concerns of the carrier's integrity and financial health. In an age of *supply chain risk management*, the carriers that a firm selects serve as a direct reflection of the hiring firm itself. Legal implications and societal perception often hold the hiring company responsible for the decisions it makes in this regard. For this reason, many shippers are electing to maintain or initiate private fleet operations, to maintain greater control over operations. When outside companies are hired for service, the shipper seeks assurances that it will regard the freight as if it were its own. The shipper therefore evaluates not only the equipment and operating capabilities of the prospective carrier, but also the commitment the shipper brings to the job.

Common Mode/Carrier Selection Criteria

Transit time reliability	Freight loss and damage
Competitive rates	Shipment expediting
Door-to-door transit time	Quality of operating personnel
Willingness to alter cost/service	Shipment tracing
	Scheduling flexibility
Financial stability	Line-haul services
Equipment availability	Claims processing
Frequency of service	Quality of salesperson
Pickup/delivery service	Special equipment

Figure 5-5   Common mode and carrier selection criteria.

The survey of different transportation modes in Chapter 2, "A Survey of Transportation Modes," acknowledged the relative strengths and weaknesses associated with each mode. The differences are not always so obvious, however. For instance, it should not be taken for granted that a shipment traveling 1,200 miles (2,000 kilometers) in 24 hours must necessarily travel by plane. A single truck driver might find this difficult to achieve while observing hours-of-service (HOS) requirements for rest, but a team of drivers working together *could* accomplish this feat. The provision of *team drivers* allows a crew of two drivers to share the driving duties. When one driver operates the truck, the other rests. This allows a truck to move virtually nonstop and greatly extends the reach by truck in a 24-hour period. Trucking companies are actively recruiting individuals to participate in team-driving arrangements. Married couples are sometimes regarded as good prospects for such arrangements, given the many hours spent together on the road!

Another development that is gaining popularity, particularly among large truck carriers, is the idea of *relay networks*. As in a track relay, in which one runner carries the baton a prescribed distance and hands off the baton to the next runner, truck relays involve handing off freight to a different driver. Truck drivers can ordinarily cover about 500 miles in a single shift. With this understanding, the trucking companies designate handoff locations approximately 500 miles apart. Figure 5-6 illustrates how one such network might appear for a truckload carrier offering service across the United States, from Baltimore, Maryland, to Ontario, California. The carrier has relay points approximately 500 miles apart on the Interstate 70 corridor, a primary east–west route for truck traffic across the center of the nation. One driver initiates the trip by collecting the freight in Baltimore and heading east to the first relay point (Columbus, Ohio, 411 miles away). Upon arriving in Columbus, a "fresh" driver collects the trailer and continues westward to St. Louis, Missouri, 417 miles away. The relay itself might involve the second driver simply assuming the tractor and trailer of the first driver or switching the trailer to a second tractor, depending on whether the tractors are dedicated to individual drivers. The second driver delivers the freight to the third relay point (Salina, Kansas), where a similar handoff occurs. This pattern repeats until the relay reaches the ultimate destination (Ontario, California). This process can happen much quicker than scheduling a single driver for the entire shipment, in light of the required rest breaks the driver must make. What makes these relay networks particularly compelling today is that drivers have more opportunity to get home more often. Traditionally, a driver might make the cross-country trip out and back alone, requiring a week or more. Under the relay provision, drivers ordinarily drive outbound one day, then rest, and then return the following day (hopefully, with a load to support a relay in the other direction). This affords the driver an opportunity to be home approximately every other day, which is much better than the tradition and seems to aid somewhat in recruiting new drivers.

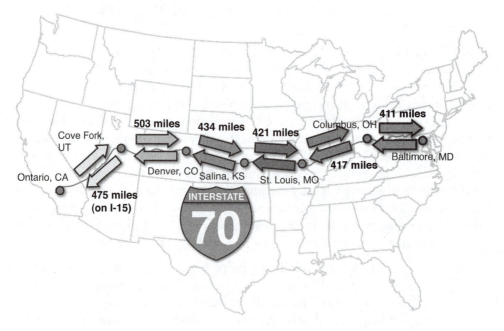

Figure 5-6   A cross-country relay network.

Under the scenarios of team drivers and relay networks, trucks can cover distances much quicker than conventionally believed. In fact, it can be argued that truck carriers using these methods can compete effectively with airplanes in providing fast, long-distance deliveries within market.[2] The same might be said of intermodal truck-rail-truck service that receives high priority among the railroads. North American railroads have devised certain corridors of their rail network that offer dedicated intermodal service, meaning that the trains move at faster speeds on the rail and have faster transloading at the origin and destination rail terminals. Under these arrangements, intermodal transportation is competing effectively with longhaul trucking and even air transportation, in light of the cost advantages of shipping by rail for the long-distance segment.

These examples speak of a "blurring among the modes," in which modes are effectively competing against one another for shares of the transportation market. Therefore, we cannot necessarily assume that a shipment should move by the same mode over time, as new service offerings enter the market from competing modes. In fact, the two best options for carrier selection might not reside within the same mode (truckload carrier A versus truckload carrier B). Instead, truckload carrier A and intermodal service provider C could be competing for the business. With this in mind, it is not always wise to assume that a given load should always move by way of the same mode.

To accommodate decision making that reviews the service offerings of different carriers operating in different modes, Figure 5-7 illustrates a method of simultaneously evaluating carriers' offerings, regardless of the mode or class in which they operate. In other words, instead of simply evaluating truckload carriers for a specific mode, carriers in other modes (such as rail, intermodal, air, or water) might be considered—so might carriers operating in other mode classes (such as LTL or parcel). The diagram suggests that, upon entering the customer service requirements for a shipment, the characteristics of products being shipped (dry versus temperature controlled, normal versus hazardous, ordinary versus high security, and so on) and the cost constraints imposed on the shipment, a database populated with the available options would be searched. The carrier that offers the best option, regardless of mode, would be selected for service. Alternatively, the method could be designed to present the five best alternatives, allowing a decision maker to select among these options. As new carriers enter the market or existing carriers alter their service arrangements and pricing, these data can be entered into the system for the most up-to-date collection of available service offerings. TMSes today can provide the support for this level of decision making. Furthermore, carriers can readily update their new offerings and prices to ensure an up-to-date database. The shipper can also enter performance data to track the carriers' performance so that these past experiences can influence future decisions regarding mode and carrier selection.

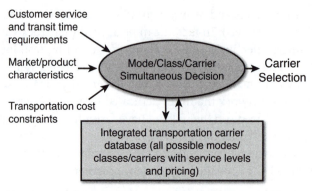

Adapted from: Stank, Theodore P. and Thomas J. Goldsby (2000), "A Framework for Transportation Decision-Making in an Integrated Supply Chain "*Supply Chain Management:An International Journal*, Vol. 5, No. 2, pp. 71-77.

Figure 5-7   Simultaneous mode/class/carrier selection.

Service Negotiations

Negotiations on matters of transportation often occur at two different levels. At the first level, the trading partners in the supply chain that sell and buy goods from one another must determine which party will be responsible for which aspects of the transportation component of the transaction. At the second level, the party responsible for hiring a carrier (when a private fleet is not used) will enter into negotiations with the carrier. This section reviews these two levels of negotiation.

Shipper–Receiver Negotiations

When a seller and buyer engage in business, they negotiate not only on matters of the products to be exchanged among the parties, but also on the means by which that exchange will happen. Central to the exchange process is the transportation of the goods. This determination is no small matter; the party that assumes responsibility for the safe delivery of the freight is accountable when something goes wrong, such as when the delivery is late or goods are lost or damaged. The seller and buyer must therefore be very clear about who is responsible for the in-transit goods and who will pay for the provision of transportation. These responsibilities and provisions are spelled out in the *terms of sale*, or *free on board* (*FOB*) terms.

Figure 5-8 illustrates the basic varieties of FOB terms. The first determination is associated with the decision of where the transaction between the seller and buyer technically occurs. The *FOB origin* designates that the sale of the goods between the two parties occurs at the seller's shipping dock. When the goods are loaded from the seller facility onto a transportation vehicle at the origin shipping point, they then belong to the buyer. Under such an arrangement, the buyer typically assumes the risk of the goods at this point, including the selection of the carrier, the risks associated with the goods when they are in transit, and payment to the carrier for the transportation service. The *FOB destination*, on the other hand, typically shifts these responsibilities and risks to the seller.

Who bears risk and who pays?

	Who assumes title and bears risk of transportation?		Who pays the carrier?		Who ultimately bears the cost of transporation?	
Terms	FOB origin	FOB destination	Prepaid	Collect	Charged back	Allowed
Responsible party	Buyer	Seller	Seller	Buyer	Buyer	Seller

Figure 5-8   Delivery terms of sale.

Further distinction on who pays the carrier and bears responsibility for the transit can be found in the payment terms of collect and prepaid. A shipment with prepaid terms means that the seller pays for the freight and incurs this cost, regardless of the FOB origin or destination determination. Inversely, *collect* terms mean that the buyer pays for the freight and incurs this cost. So under *FOB Origin, Collect* terms, the buyer assumes full responsibility, risks, and cost associated with the shipment. Under *FOB Origin, Prepaid* terms, however, the transaction for the goods occurs at the seller location; the buyer bears the risk of the freight, yet the seller pays the carrier and assumes the cost of the transportation. Buyers and sellers typically enter into FOB Origin, Collect (where the buyer assumes full responsibility) or *FOB Destination, Prepaid* (where the seller assumes full responsibility) to avoid the confounds between who selects the carrier, pays for the service, and bears the risk associated with the service. FOB Origin, Prepaid and *FOB Destination, Collect* remain viable options, however.

A final twist on the FOB terms of delivery is found in the provisions of Charged Back and Allowed. *Charged Back* is a provision that can be added to the FOB Origin, Prepaid arrangement. In the absence of the Charged Back provision, the seller pays for the freight. With Charged Back, the seller still pays for the freight but invoices the buyer in the amount of this service, or charges back for the arrangement. As for *Allowed* terms, this provision permits the seller to deduct the cost of the transportation service from the amount it pays the seller under FOB Destination, Collect terms. Figure 5-8 helps to provide clarity on these important distinctions.

Shipper–Carrier Negotiations

In a free market environment, shippers and carriers are permitted to enter into negotiations on the specific services to be provided, the assurances and penalties that will be associated with the service, and the price for the service. Chapter 3, "The Economics of Transportation," reviewed the different pricing parameters for service and the ways in which transportation rates are expressed. Here we incorporate the service and relational aspects of the negotiation. We start by examining the nature and expected duration of the relationship between the shipper and the carrier. This manifests in whether the shipper enters into a contractual relationship with the carrier or, instead, elects to engage on a transactional basis. Whereas contractual arrangements outline the service expectations, rates, and relational aspects of the business between the two companies over several transactions that might span a year or more, transactional arrangements focus on a single transaction between the two firms. As expected, negotiations for a long-term arrangement typically involve more preparation and involved discussions, although one cannot neglect the customer service implications or liability potential in individual transactions.

Contract Versus Spot Rates

Shippers buy transportation services from carriers under contract and spot rates. *Contract rates* involve agreed-upon prices for services between a shipper and a carrier for a specified period of time, usually one year. *Spot rates* are market prices offered for services on a specific transaction. It is estimated that 80 percent of freight moves under contract in the United States. Shippers choose to enter into contracts as a means of locking in prices for an extended time period, gaining the commitment of the carrier to offer capacity and agreed-upon service levels. Contracts spell out the commitment between the parties and often carry penalties for noncompliance applicable to both parties in the transaction.

Large shippers seek aggressive price reductions over spot prices that are offered to any prospective buyers. As in any industrial buying situation, buyers with significant influence in the market can garner deeper discounts based on their buying power. Transportation is not different in a free market environment. Large shippers are also more likely to have private fleets or to have the financial resources to devise private fleets. The ability to leverage a private fleet in negotiations with carriers exerts additional downward pressure on prices. Conversely, small shippers lack the buying power and influence in their negotiations with carriers. This sentiment has incited many small shippers to encourage re-regulation of markets, in which prices are fixed for small and large shippers. Agricultural shippers, in particular, have been impacted by losses of service and rate increases for service that remains in low-density areas of population. These shippers were protected in the United States in the days of the Interstate Commerce Commission (ICC), as well as in the state rate bureaus for intrastate transportation.

Contractual Provisions

When a shipper elects to hire one or more carriers on a contractual basis, it can involve an extensive shopping experience. Firms sometimes engage in *request for proposal (RFP)* events as a way to capture the ideas of prospective service providers and their prices. An RFP usually involves inviting select carriers to review the business of the hiring company, including the existing supply chain operations, shipping lanes, and volumes. Service providers are encouraged to devise solutions that best meet the needs of the shipper. Conversely, some shippers issue *requests for quotes (RFQs)*, which spell out the specific service arrangements that the company is seeking. Instead of looking for broad solutions, the hiring company is seeking the best prices and commitments to perform specific transportation and logistics services.

Under both RFP and RFQ arrangements, the competition is usually blind, meaning that competitors are operating independently, without knowing the provisions or pricing of competing bids. The competition often involves two or more rounds as the shipper narrows the carrier(s) it will hire for the service. Carriers are encouraged to "step up" their bids at each iteration, elevating service performance and lowering price along the way. Many large shippers engage in RFQ (and even RFP) events annually. Such frequency allows for adaptations to be made each year in the service arrangement. It can also allow the carrier to adjust pricing—either downward to reflect the benefit of learning the shipper's business and to reflect any economic deflation that might occur in the market, or upward to reflect the realities of serving the customer or any inflationary forces at work in the economy. For large projects, shippers and service providers commonly enter into multiyear arrangements (three- or five-year deals are common), with provisions built into the agreement to allow for adjustments in service and pricing as the respective parties allow.

The buying power of the customer often influences how much persuasion it has in the market. Large firms are usually able to enjoy more competitive prices and higher priority service in light of the large volumes they ship and the account size they represent. Smaller shippers must therefore appeal to carriers on different grounds. These might include the types of products they ship (for example, easy to handle, well contained, nonhazardous), the lanes over which they operate (for example, presenting backhaul opportunities), or the time of year in which they ship (for example, shipping in off-peak seasons when carriers are in need of freight). Another hard-to-quantify way of appealing to carriers is by being easy to do business with. When companies conduct business, it ultimately involves people working together across organizations. All businesses must ensure profitability to survive in the long run, but there is much to be said for organizations that take an interest in the success of a partnering company. Small firms can sometimes appeal to carriers on this basis. This can manifest in several different ways, such as providing ample time for dispatches, performing loads and unloads promptly to minimize waiting, treating the carrier's employees (such as drivers) with respect, and paying freight invoices promptly. These seemingly minor courtesies can elevate the status of shippers, large or small, in the eyes of carriers. It is often easier for smaller shippers to affect these kinds of attitudes and actions because of their smaller scope of operations and ability to instill culture among a more limited set of people.

When shippers and carriers agree to enter into negotiations, attention turns to the specific services, terms, and prices. Among the service aspects considered are the speed, volume, frequency of shipments, and reliability of service. Shippers will be concerned with the service provider's track record in these regards, seeking assurances that the carrier has the capabilities and capacities to accommodate the need. Also of significant importance is the assumption of liability, a clear understanding of the responsibilities of each party, and any limits to over, loss, and damage (OS&D) claims. Shippers will also be concerned with the accommodation of any special needs they might have, whether operational in nature (for example, security or temperature control) or administrative (such as with documentation requirements). When attention turns to the price for service, focus is directed to the charges for basic line-haul transportation services, as well as any surcharges and accessorial fees that might apply.

Finally, the length and nature of the contract is determined. For reasons noted earlier, some shipper–carrier relations involve multiyear contracts. This is particularly true when the shipper is seeking to devise a *core carrier program*. Such an arrangement identifies a select group of strategic carriers that will receive a large share of the shipper's volume. The idea is to concentrate these volumes in the hands of fewer carriers, to enjoy larger discounts in exchange for awarding higher volumes of freight to these chosen carriers. The shipper also expects to receive higher priority service from the core carriers. This is particularly valuable when capacity is tight and shippers are competing for the interests of the carriers. The risks of core carrier programs include the lost freedom of shopping for lower prices in market, in light of the volume commitments made to the carriers. Another risk is the prospect of misplacing trust in the hands of carriers that cannot or choose not to fulfill the service obligations, or the risk of a core carrier weakening operationally and financially over time, with no ready substitutes in place. Core carrier programs can usually accommodate these challenges, however, through careful selection and monitoring of the carriers.

Other times, relationships might enter into annual contracts, or perhaps project-specific agreements of shorter durations. When a shipper closes a manufacturing plant or distribution center, for instance, it might contract with a carrier or 3PL to move all the freight from the closed facility to another location. Depending on the size of the facility and the volume of goods housed there, such a transfer might take a few days or several months. Along with the contract duration, the two parties will agree on any special conditions that either party might impose on the relationship. Adjustments in service, pricing, or payment terms might be among these conditions. Finally, the two parties will agree on the terms of any early termination of the contract that might be deemed necessary. The provision of having two parties enter into new business together can be a joyous one (as in a new marriage), but the two parties must also determine what would merit a dissolution of the arrangement and how any such dissolution should be handled. Although it is impossible to imagine every possible circumstance, the two parties should rely on their respective experiences to outline service, economic, and relational grievances considered unacceptable in the arrangement. On a more positive note, the two parties might also spell out the rewards that might accrue to the parties if performance exceeds expectations. Such concerns are iterated in the next section on service evaluation.

Service Evaluation

In any activity, business or otherwise, it is important to measure performance. Several popular expressions underscore this premise: "If you don't keep score, you're only practicing"; "What gets measured, gets managed"; and "You get what you inspect, not what you expect." These sentiments find application in the management of transportation services. Measuring performance across the various aspects of the service arrangement is required to ensure that the service is living up to expectations. However, the challenge of identifying *which* measures to employ is critical because transportation is such a broad and encompassing business activity.

The broad dimensions of performance to consider include safety, service performance, cost, and relational performance. To the extent possible, shippers should devise metrics that can be tracked on these various dimensions. To the extent that the metrics are quantitative and verifiable (for example, on-time performance, damage rate, and price variation), the shipper and carrier can better agree on quality of performance. Other metrics tend to be more qualitative for service aspects, as with ease of doing business. Complaints can be measured in terms of frequency, for example, but the nature and severity of issues can be difficult to express in numerical terms. That said, efforts should be made to express even these qualitative concerns in simple, numerical means through subjective scaling. However, even seemingly tangible metrics such as on-time delivery percentage can be challenging when the metric lacks definition. In one example, a customer claimed that a shipper was routinely delivering product late to the customer's distribution center. The shipper disagreed. Eventually, the two parties agreed to watch a truck deliver the shipper's goods. It was discovered that the shipper considered the delivery complete when the truck arrived in the receiving yard of the distribution center. The customer, however, considered the delivery complete when the truck was dispatched from the receiving yard and backed up to the dock for unloading. This simple difference of interpretation resulted in very different views of the shipper's delivery performance. Clarity in the metrics is essential if they are to be used to manage the business.

To keep the measurement system manageable, it is imperative to prioritize the number of measurements used to assess performance. In other words, less meaningful and significant measures need not factor significantly into an evaluation. In fact, many companies should consider eliminating metrics that are no longer valuable but that still require time to collect and report. In some cases, the metrics can even be detrimental in making properly informed decisions because they are distracting, invalidated, or misguided in light of new business priorities. Shippers and carriers alike are encouraged to question the measures they use and to adapt the measurement system from time to time to reflect the changes occurring in the business.

Figure 5-9 illustrates a sample scorecard that a shipper might use to evaluate the performance of a carrier. Such a scorecard is valuable not only for internal purposes at the shipper, to evaluate the comparative performance of different carriers, but also to communicate directly to the carrier. Scorecards of this nature are quite common in evaluating material suppliers—increasingly so in evaluating service providers such as transportation companies. A scorecard such as the one in Figure 5-9 might be gathered monthly and shared with the carrier to communicate the perceived service received.

Performance Criteria	Maximum Score	Carrier Score	Comments
Transit time consistency	20	18	Weather aside, no big problems
Meets delivery	16	14	Missed 4 windows this month
Equipment availability	14	8	Too few refrigerated units
Rates competitive	12	10	Should renegotiate FAK rate
Emergency response	10	6	Costly dispatch error (9/21)
Incident-free performance	8	6	Minor dock damage (9/3)
Driver courtesy	6	5	No real problems
Ease of doing business	5	3	Can never talk to same person
Billing accuracy	5	4	Six (minor) errors in 96 bills
Claims settlement	4	4	Two claims cleared this month
TOTAL SCORE	**100**	**78**	Need to resolve people and equipment issues; service is fine on scheduled runs

Figure 5-9   Sample scorecard for carrier performance.

Note that the scorecard contains ten different performance criteria, and the criteria have different weights attached (by having different maximum scores). Transit time consistency is the single most important criterion in the example, signifying the importance for the shipper that the carrier be very consistent in completing transit times against expected time parameters. The carrier scores listed in the next column, then, refer to a conversion of data and observations collected over the past month regarding performance on each criterion. The next column provides a succinct statement justifying the score that was issued.[3] Furthermore, the shipper might institute rules associated with the score. Perhaps scores below 80 on the 100-point scale might justify a face-to-face meeting with the carrier to determine ways to remedy the issues that are bringing down the score. In the current example, it appears that the availability of equipment and response to emergency situations are proving particularly troublesome for the shipper, and root cause analysis might be necessary to determine the reasons and remedies for these issues.

To the opposite effect, many shippers and carriers are embracing *performance-based logistics (PBL)* programs that reward carriers for excellent service. The rewards might include commitments of future business or financial rewards. The U.S. Department of Defense is particularly active in devising PBL relationships with its logistics service providers, and the trend is growing. Beyond these formal arrangements, however, other shippers find great value in simply recognizing carriers for excellent service by way of *Carrier of the Year* awards. These can be low-cost ways of acknowledging carriers for excellence that benefit the carriers in multiple ways. For one, the carrier likely is in an improved position for future business with the focal company. For another, the carrier can use the award as an endorsement of valuable service, to gain new business. Any such recognition should be based on the routine and discipline measurement of the key carriers a shipper employs.

This discussion focuses on the shipper's measurement of carrier performance, but it is important to note that carriers should also evaluate the performance of shippers. They need to ask, "Are our customers living up to *their* expectations?" These expectations include several of the same dimensions outlined previously for carriers, including safety, ease of doing business, and economics (payment). It is not uncommon today for carriers to maintain scorecards of shipping customers and to critically evaluate the performance of each key customer. Such evaluation is warranted when shippers are vying for the attention of high-performing carriers, especially in a capacity-constrained market. Shippers under these circumstances who are seeking to be recognized as a "customer of choice" or a "preferred customer" achieve priority when the supply of service is limited in some way.

Dock- and Movement-Level Decisions

The final level of decision making in transportation occurs with each shipment. These decisions usually occur at the site of the work—at the shipping and receiving docks—as well as in the distance that separates the two. Here an intense focus on process is required, to examine *what* work gets performed and *how* it is performed. It is critical to realize, too, that all the planning and big ideas are executed at this level. The best plans are worth little if the processes associated with execution are ill equipped to handle them. For this reason, many companies are employing continuous improvement methods, such as lean and six sigma, in the design of work processes. *Lean* emphasizes the elimination of waste, and *six sigma* focuses on variation in processes. They are often employed together, for where there is variation in process, one will also find waste. Hence, *lean six sigma* is finding good application in the operational and administrative processes occurring in dock- and movement-level activities. Tools such as *value stream mapping* and *process mapping* can identify and evaluate defects in work processes.

Instead of focusing on the minutia of how to process freight on the dock or how to operate transportation equipment, we focus here on the information required to make informed decisions in support of dock and movement activity. From the use of *routing guides* that inform a shipping clerk which carriers to consider on a particular lane, to in-route instructions for safely navigating a transportation vehicle to the rightful destination, information is vital for effective and efficient performance of work at the operational level.

However, information is not the only vital element in the effective conduct of work at the operational level—hiring, development, and retention of the workforce is also important. More than 80 percent of all logistics activity occurs beyond the view of supervision. Virtually all transportation activity lacks direct supervision, although rapidly advancing technologies such as global positioning satellites and equipment-monitoring devices are changing this tradition. Despite these advances, companies must hire and develop people competent for the work on hand, train them to do the work safely and successfully, and retain this essential talent for continuity. These general principles hold true for shippers and service providers alike.

Although the operational aspects of transportation (shipping, transporting, receiving) likely come to mind, effective transportation management relies extensively on administrative activities that support the operations as well. Certain administrative tasks that occur before, during, and after any transportation movement require high levels of performance for any transaction to ultimately be deemed successful. Much of this activity focuses on the accurate completion and sharing of transportation documentation. Table 5-2 illustrates essential transportation support activities that require documentation. Figure 5-10 shows a sample of the information exchange present in a typical transportation transaction among a shipper, a carrier, and a receiver.

Table 5-2   Essential Transportation Support Activities

Transportation Activity	Information User		
	Shipper	Carrier	Receiver
Pretransaction	Purchase order information	Bill of lading information	Advanced ship notice
	Forecasts	Forecasts	Delivery time
	Equipment availability	Pickup and delivery time	
Transaction	Shipment status	Shipment status	Shipment status
Posttransaction	Freight bill	Payment	Carrier performance
	Carrier performance	Claims information	Proof of delivery
	Proof of delivery		Claims information
	Claims information		

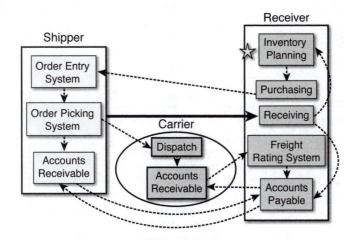

Figure 5-10   Typical information exchange in a transportation transaction.

Transportation Documentation

Appropriate documentation is an extremely critical (and often overlooked) aspect of transportation. Documentation needs vary based on the nature of the shipment (for example, international versus domestic), nature of the cargo (for example, normal versus hazmat versus perishable), and trading partners (for example, exporting to and importing from some countries require special documentation). This is why logisticians have a saying: "Freight moves on a sea of paper!" Given the complexity of the process, it comes as no surprise that both shippers and receivers must insist on absolute clarity in documentation; any mismatch can be an extreme source of frustration to both sellers and buyers.

Documentation time and resources usually account for anywhere between 5 and 10 percent of the total value of all shipments. (Several of these documents can now be transmitted electronically, thereby reducing the paperwork burden somewhat.) In this section, we cover several of the documentation requirements of transportation (see Figure 5-11). We begin with the documentation that is common to all forms of transport (domestic and international) and then cover some specific documents that are required in special cases.

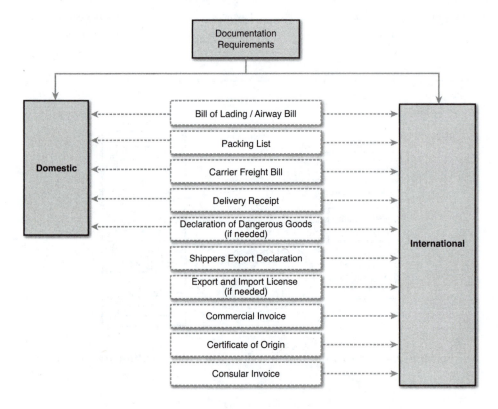

Figure 5-11   Transportation documentation requirements.

Documents Common to Domestic and International Transportation

Some documents are common to domestic and international transportation and will be encountered regardless of whether the freight crosses international borders:

1. **Bill of lading (B/L or BOL)**—This is possibly the single most important document in logistics (both domestic and international). The *bill of lading* is a shipping

document that a freight carrier issues, acknowledging that certain named goods have been received onboard as cargo for conveyance to a named place, for delivery to the consignee. (3PLs and freight forwarders are also authorized to issue bills of lading, but freight brokers are not.) Among other points, the bill of lading serves the following purposes:

- It is evidence of a contract of carriage between ocean freight carrier and shipper.

- It is a receipt for goods.

- It is a document of title on shipped goods.

The bill of lading also contains the following information:

a. Name of the shipping company and its registration

b. Shipper's name

c. Order and notify party

d. Description of goods

e. Gross/net/tare weight

f. Freight rate/measurements and weight of goods/total freight

Depending on the negotiations between the buyer and seller of goods, the B/L can be created either as a negotiable or a non-negotiable instrument. A *negotiable bill of lading* (also called *order B/L*) transfers ownership rights of the named goods to anyone who has possession of the instrument (B/L).[4] Therefore, the negotiable B/L can basically be traded or sold to someone in exchange for things of value (called *factoring*). A negotiable B/L has the words "To Order" written on it, and the original is usually required to take delivery of the shipped goods. A *non-negotiable B/L* (also called a *straight B/L*) cannot be traded and has no monetary value, except that it also confers ownership of goods. As such, if goods are shipped under a straight B/L, the original B/L is not needed to take possession of the goods. When goods are shipped by way of air freight instead of ocean freight, the bill of lading is called an *airway bill*.

2. **Packing list**—A packing list is a document that the shipper prepares, listing the kinds and quantities of merchandise in a particular shipment. It also has detailed information on the type, size, and weight of each container in the shipment. A copy of the packing list is often attached to the shipment in a waterproof envelope, and another copy is sent directly to the consignee to assist in checking the shipment when received. It is also called a *bill of parcels*. In international shipping, it is absolutely imperative that the packing list match the commercial invoice (discussed later), or the shipment can be delayed through Customs, perhaps for months.

3. **Carrier freight bill (CFB)**—The CFB is an invoice presented by the carrier to the shipper, the consignee, or a referenced third party as a demand for payment for services rendered. Similar to the B/L, it is a standard document and shows the name of the carrier, the carrier's reference number/Standard Carrier Alpha Code (SCAC), the shipper's name and address, consignee's name and address, a description of the goods, the rate, freight terms, and the charges due.

4. **Delivery receipt**—The delivery receipt is a document issued by the carrier that the consignee signs as proof of receipt of the shipment. It is also known as a *proof of delivery (POD)* document. The carrier and the consignee each retain a copy of the delivery receipt.

5. **Declaration of dangerous goods (if needed)**—A shipper's declaration of dangerous goods is a letter that describes the dangerous goods (hazardous materials) and quantity shipped and provides key information to communicate hazards present for safe transport and for mitigating spills or leaks. The letter must also include this information:

   a. Total quantity of the materials

   b. Technical description for generic shipping names

   c. Emergency phone number (that is answered 24/7 by a knowledgeable person)

Documents Exclusive to International Transportation

A second (and more extensive) set of documents is almost exclusively required when freight crosses international borders. These documents are in addition to the ones that we have already discussed:

1. **Shipper's Export Declaration (SED)/Automated Export System (AES)**—The SED filing is generally required by the U.S. Census Bureau for U.S. exports when a single commodity's value exceeds US $2,500, or when a postal shipment's value exceeds US $500. The SEDs must also be prepared, regardless of value, for all shipments requiring an export license or for shipments to embargoed countries (such as Zimbabwe, Cuba, Iraq, Sudan, Syria, and North Korea). The SED is used for two purposes:

   ■ It serves as a census record of U.S. exports. The government generates many reports using these statistics.

   ■ It serves as a regulatory document.

   Earlier, a paper-based document had to be filed to comply with this requirement, but a shipper can now file it electronically with the AES using the AESDirect website (www.aesdirect.census.gov) of the U.S. Census Bureau. After the export declaration is successfully filed and processed, the shipper receives an Internal Transaction Number (ITN) to put on the shipping documents, as confirmation for any government agent inspecting the cargo before departure.

2. **Export and import licenses (if needed)**—Most export transactions do not require specific approval in the form of licenses from the U.S. government, but some do. It is typically up to the shipper to determine whether the product requires a license and to research the end use of the product—in other words, to perform "due diligence" regarding the transaction. Exporters should learn which federal department or agency has jurisdiction over the item they are planning to export so that they can find out whether a license is required. Similarly, an import license is a document that gives the buyer the permission to import goods into the country. In general, there is no need for an import license for importing goods into the United States. However, there are some restrictions on goods such as alcohol, tobacco, firearms, animals, copyrighted materials, food, and artifacts. Note that many other countries do require the buyer to have a license to import goods into the country.

3. **Commercial invoice**—A commercial invoice is used when dutiable goods are shipped internationally; it works as a sort of Customs declaration. The invoice should be completed on the shipper's company stationery, must contain the shipper's complete company address and telephone number, and must be signed by the shipper or its agent. An accurate and complete description of goods is necessary for Customs purposes. If the receiver/consignee is different from the importer/buyer, the invoice should note that. Three copies of this document are usually created: the original commercial invoice, one copy attached to the shipment paperwork, and one copy attached to the actual shipment.

4. **Certificate of origin (CO)**—A CO is a document that states the country where the shipped goods originated. Note that the term *originate* in a CO does not mean the country the goods are shipped from, but rather the country where the goods are actually made. In general, as long as more than 50 percent of the value of the goods originates from a country, that country is acceptable as the country of origin. The CO is useful for classifying the goods in the Customs regulations of the importing country, thus defining how much duty shall be paid. Moreover, it might also be important for import quotas and for statistical purposes.

5. **Consular invoice**—In general, the term *consularization* refers to the practice of getting any document approved by the consulate in another country. Usually, consularized documents have a red seal or ribbon on them to mark as such. A commercial invoice that has been consularized is called a consular invoice. Consular invoices have been around for a long time and were supposed to ease the flow of trade (by easing flow through importing Customs), but their popularity is waning and several commodities no longer need to be consularized.

Two areas of transportation management require special consideration and often additional training to gain expertise. These areas are the handling of hazardous materials and international transportation, in light of their inherent risks and level of expertise required to ensure safe, efficient transit of goods. As noted, the accuracy of documentation associated with all transportation is essential, but that is especially true for international movements.

As with many activities in transportation and business in general, technology improves the ease and efficiency of information exchange. Furthermore, information technology (IT) provides decision support assistance to improve the speed and quality of decisions. Figure 5-12 illustrates some of the different ways information technologies can influence and inform decisions ranging from very macro to micro in orientation. As shown, IT can play a key role in all levels of the decision-making framework, from informing strategic analyses, such as network design and forming consolidated shipments, to making decisions at the shipping dock, such as those related to load building and sequencing.

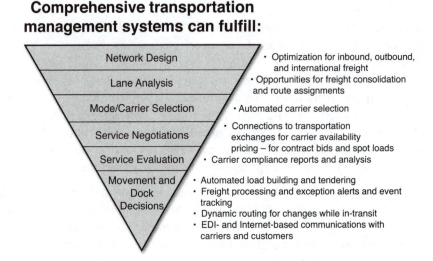

Figure 5-12   Technology support for transportation decision making.

Summary

The next chapter offers a more in-depth look at TMS and related technologies that help reduce the uncertainties and inform transportation decision making.

Key takeaways from this chapter include:

- Transportation management affects many different aspects of the focal company, as well as its interactions with suppliers and customers. Transportation must therefore be managed on an integrated, holistic basis with other functions.

- Transportation management decisions should flow from the most strategic level to the operational level, to ensure that the business priorities are met in the most effective manner.

- Strategic decisions guide operational actions, and operational capabilities influence the strategies available to the company.

- Decisions and actions taken at the operational level ultimately determine the effectiveness of a company's transportation management.

- It is essential that documentation supporting transportation activity be accurate and timely, to ensure that the full value of operations is rendered in delivery.

Endnotes

1. Students have identified more than 20 different forms of load consolidation and class shifts in these shipping data. Granted, some consolidations could have questionable implications for service, given the number and timing of stops involved.

2. The limitation of truck transportation, of course, is its inability to travel over water without the aid of intermodal or roll-on/roll-off support. Hence, we limit this comparison to shipments within a single landmass that do not involve an overseas component.

3. More justification is required than the few words used here merely for illustration.

4. The negotiable/order bill of lading is rarely seen in domestic transportation.

6

TRANSPORTATION TECHNOLOGIES

By definition, a supply chain includes the flow of information (along with material) to and from all participating entities. Many, if not most, of the supply chain problems are the result of poor flow of information, inaccurate information, untimely information, and so on. Information must be managed properly in each supply chain segment. Information systems are the *links* that enable communication and collaboration along the supply chain. They represent one of the fundamental elements that link the organizations of supply chain into a unified and coordinated system. In the current competitive climate, little doubt remains about the importance of information and information technology (IT) to the ultimate success, and perhaps even the survival, of any supply chain management (SCM) initiative. According to a recent survey of 2,500 large and small companies, SCM was the largest focus area for future IT investment (as compared to sales and operations). As Figure 6-1 illustrates, more executives suggest that IT associated with SCM will see either "Much Investment" or "Maximum Investment" in the coming years, as compared to either sales or operations management-related IT.

Case studies of some world-class companies, such as Walmart, Dell Computers, and Federal Express, indicate that these companies have created sophisticated information systems that exploit the latest technological developments and create innovative solutions. Table 6-1 shows representative IT solutions, together with the problems they solve.

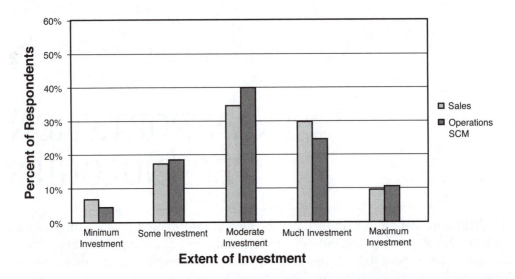

Figure 6-1   Extent of future planned investment in key business functions.

Table 6-1   IT Solutions to SCM Problems

Supply Chain Problem	IT Solution
Slow linear sequence of processing	Employ parallel processing, using workflow software.
Excessive wait times between chain segments	Identify the reason (Decision Support System software) and expedite communication and collaboration (intranets, groupware).
Existence of non-value-added activities	Conduct value analysis (SCM software); use simulation software.
Slow delivery of paper documents	Use electronic documents and an electronic communication system (such as electronic data interchange [EDI] or email).
Repeated process activities due to wrong shipments, poor quality, and so on	Use electronic verifications (software agents) and automation; eliminate human errors with electronic control systems.
Batching; accumulated work orders between supply chain processes to get economies of scale (for example, to save on delivery)	Use SCM software analysis; digitize documents for online delivery.
Delays identified after they occur or identified too late	Use tracking systems, anticipate delays, conduct trend analysis, and employ early detection (intelligent systems).

Supply Chain Problem	IT Solution
Excessive administrative controls, such as approvals (signatures); approvers in different locations	Use parallel approvals (workflow) and an electronic approval system. Analyze need.
Lack of information, or too-slow flow	Use the Internet/intranet; use software agents for monitoring and alert. Employ bar codes and direct flow from point-of-sale (POS) terminals.
Lack of synchronization of moving materials	Use workflow and tracking systems. Employ synchronization by software agents.
Poor coordination, cooperation, and communication	Use groupware products, constant monitoring, alerts, and collaboration tools.
Delays in shipments from warehouses	Use robots in warehouses; use warehouse management software.
Redundancies in the supply chain; too many purchasing orders, too much handling and packaging	Share information via the web, creating teams of collaborative partners supported by IT.
Obsolescence of parts and components that stay too long in storage	Reduce inventory levels by sharing information internally and externally, using intranets and groupware.

On the transportation front, one of the most important technology-related topics is information sharing along the transportation channel (and, furthermore, along the entire supply chain). The primary reason (and this is by no means the only reason) for having IT in the transportation channel is to improve communication and decrease complications such as order mismanagement, improper forecasting, improper ordering, and inefficient vehicle routing. To understand this better, the next section explores a phenomenon central to supply chains: the bullwhip effect.

Understanding the Need for Technology: The Bullwhip Effect

The *bullwhip effect* refers to erratic shifts in orders up and down the supply chain. Procter & Gamble (P&G) initially observed this effect in the company's disposable diapers product, Pampers. Although actual sales in stores were fairly stable and predictable, orders from distributors had wild swings, creating production and inventory problems for P&G. An investigation revealed that distributors' orders were fluctuating because of poor demand forecast, price fluctuation, order batching, and rationing within the supply chain. All this resulted in unnecessary inventories in various locations, fluctuations of P&G orders to suppliers, and flow of inaccurate information. Distorted information can lead to tremendous inefficiencies, excessive inventories, poor customer service, lost revenues, and missed transport schedules.

The bullwhip effect is not unique to P&G. Firms from Hewlett-Packard in the computer industry to Bristol-Myers Squibb in the pharmaceutical field have experienced a similar phenomenon. Basically, even slight demand uncertainties become magnified when viewed through the eyes of managers at each link in the channel. If each distinct entity makes ordering and inventory decisions with an eye to its own interest above those of the chain, stockpiling might be simultaneously occurring at as many as seven or eight places across the supply chain; in some cases, this leads to as many as 100 days of inventory—waiting "just in case."

Thus, information sharing among business partners, as well as among the various units inside each organization, is necessary for successful SCM. IT must be designed so that information sharing becomes easy. One of the most notable examples of information sharing is between P&G and Walmart. Walmart provides P&G access to sale information of every item P&G makes for Walmart. P&G daily collects this information from every Walmart store. By monitoring the inventory level of each P&G item in every store, P&G knows when the inventories fall below the threshold that trigger a shipment. All this is done automatically. The benefit for P&G is accurate demand information. P&G has similar agreements with other major retailers. Thus, P&G can plan production more accurately, avoiding the bullwhip effect. A 1998 industry study projected that $30 billion in savings could materialize in the grocery industry supply chains alone using such bullwhip reduction approaches. Indeed, when each part of the supply chain obtains real-time information about actual end demand, and when inventory management decisions are coordinated, inventory levels (and, consequently, costs) are reduced across the supply chain.

Thus, companies might be able to avoid the "sting of the bullwhip" through information sharing. But exactly what kind of information is this? Demand forecasts, point of sale, capacity, production plans, promotion plans, and customer forecasts are some of the many forms this information can take. A key part of this information relates to the identification and visibility of items within the transport channel, their locations, and the use of this information to make better vehicle routing and load-balancing decisions. All these help reduce distortions and errors and encourage better decision making in the supply chain.

These developments are made possible in the transport channel through technologies such as electronic data interchange (EDI), transportation management systems, routing and scheduling systems, automatic identification (bar coding and radio frequency identification [RFID]), and control and monitoring systems (such as location-monitoring systems and temperature-monitoring systems). Some of these are primarily software-driven technologies (such as EDI, transportation management systems, and, to some extent, routing and scheduling systems); others are a combination of hardware- and software-driven technologies (for example, automatic identification and control and monitoring systems).

In this chapter, we discuss many of these technologies—but first we must understand a key element of technology architecture: hosted systems (that is, locally hosted and application service provider [ASP]) versus software as a service (SaaS). A general understanding of both these concepts and the differences between them is critical to understanding the workings of all these systems.

Technology Architecture

Modern firms have two primary ways of constructing their technology architecture: by hosting their own servers and systems, or acquiring technology from outside providers.

Hosted Systems/Hosted Software

Hosted software typically implies that the user directly buys a software solution/application from a publisher or a vendor. More important, the buyer has the software installed at a data center or "hosting center," where either physical or virtualized servers are available. Typically, these servers are owned, leased, or financed. In addition, the hosting center can be either local or long distance. When the hosting center is local, the data and software are stored onsite; with long-distance hosting/ASP models, data is stored offsite. The buyer then implements the solution and uses it in the business environment.

With the payment stream, the buyer typically has a larger upfront software payment, a price for hourly or project-based implementation, possibly an initial provisioning fee from the hosting center, and then a monthly fee for the rental/usage of the hosting center's equipment, people, and bandwidth. The long-term ongoing fees include the monthly hosting fee, an annual software maintenance fee that covers bug fixes and new versions, and any hourly billed or annual contracted phone support from the vendor or publisher. Finally, the buyer might have a cost every few years for the vendor to upgrade the software to the latest version. If the software is hosted onsite, some of these costs might be eliminated (such as the monthly fees for renting the center's equipment and people). The biggest challenge with such systems is that changes and upgrades to the system are typically harder to do (because they need to be updated on various servers running in parallel). In addition, upgrades are often "patchy," in the sense that they tend to be rolled out in "blocks" instead of on a continuous basis.

Software as a Service (SaaS)

In contrast to hosted systems, SaaS applications are typically "multitenant," meaning that they serve several customers on a single software installation and database infrastructure. This means that one database shares multiple end-user customers who are "partitioned" from each other by one or more security models in the application. As a result, the initial installation and procurement are either reduced or completely eliminated. Thus, SaaS is almost always a pure web/HTML-based solution that is typically sold on a rental model, typically X dollars per month, per user. This also means that most users can access these applications with an Internet browser (such as Microsoft Internet Explorer or Mozilla Firefox), and the initial costs are typically lower. Applications designed this way are also relatively easier to scale up or down, easier to manage by the host, and easier to make self-configurable by customers. All other things being equal, this combination typically makes SaaS applications more affordable to the buyer especially for smaller applications.

In the long run, SaaS solutions can sometimes turn out to be more expensive, especially if the need for scaling up is large, or if the number of users increases substantially. However, despite these problems, buyers often prefer the cash flow management advantages that true SaaS solutions provide, even if they turn out to be more expensive in the long term. For example, from a cash flow standpoint, a buyer might find it more acceptable to implement a software application that charges a flat rate of $250 per user per month, as compared to paying $100,000 upfront for procurement and setup (even though, in the long run, the format might end up costing substantially more). Table 6-2 shows a quick comparison of the three models we discussed.

Table 6-2 Comparison of Various Technology Architectures

	Locally Hosted	ASP	SaaS
Initial cost	High	Low to moderate	Low
Long-term cost	Low to moderate	Moderate	Moderate to high
Scalability	Low to moderate	High	High
Security	Moderate to high	Moderate to high	Moderate
Technical expertise needed	High	Moderate to low	Low

Electronic Data Interchange (EDI)

In order to facilitate the timely and accurate exchange of information across organizational boundaries, many firms are turning to EDI. But, much about EDI remains to be learned.

What Is EDI?

The formal definition of EDI is "the electronic exchange of business documentation and information in a standardized format between computers, usually of different organizations." It is also commonly known as *electronic trading*.

As such, EDI is a concept or system of at least two trading partners, a computer system and a communication network. Such interorganizational computer networks support the exchange of computer-stored information across organizational boundaries. In such an arrangement, business documents such as purchase orders and invoices are exchanged (EDI messages) electronically. Examples of common EDI transactions that substitute for conventional preprinted business forms include purchase orders, materials forecasts, and shipment and billing notices. To date, the Data Interchange Standards Association (DISA) has cataloged standards for 245 transaction sets, or EDI applications. Many of these transactions sets are related to general-purpose business exchanges, but others are industry-specific (transportation, retail, or healthcare industries).

EDI replaces human-readable, paper, or electronic documents with machine-readable, electronically coded documents. With EDI, the sending computer creates the message and the receiving computer interprets the message without any human involvement. One of the first places many companies implement EDI is in the exchange of a purchase order (PO). In the traditional method of processing a PO, a buyer or purchasing agent goes through a fairly standard procedure to create a PO:

1. A buyer reviews data from an inventory or planning system.

2. The buyer enters data into a screen in the purchasing system to create a PO.

3. The buyer waits for the PO to be printed, usually on a special form.

4. After the PO is printed, the buyer mails it to the vendor.

5. The vendor receives the PO and posts it in the order entry system.

6. The buyer calls the vendor periodically to determine whether the PO has been received and processed.

When you add up the internal processing time required by the sender and receiver, and then add in a couple days in the mail, this process normally takes between three and five days. This assumes first that both the sender and the receiver handled the PO quickly and then, at every point along the way, that no errors occurred in transcribing data from a form to a system.

Now consider the same document exchange when a company places its purchase orders electronically using EDI:

1. The buyer reviews the data and creates the PO, but does not print it.

2. EDI software creates an electronic version of the PO and transmits it automatically to the sender within minutes.

3. The vendor's order entry system receives the PO and updates the system immediately upon receipt.

What took up to five days with paper and the postal system now takes less than an hour. By eliminating the paper-handling from most stages of the process, EDI has the potential to transform a traditional paper-based supply chain business process.

Benefits and Applications of EDI

Speed: Speed, whether in the increased velocity of moving products from design to the marketplace, or in the rapid response of a supplier to customer demands, is vital to success. Increased speed can benefit a business in several ways:

- Shorten lead times for product enhancement or new product delivery. The market advantage of months or even weeks can have a major impact on profitability.

- Do more with less. Staff reductions, which are common in many businesses, require that fewer people accomplish more work. Handling the exchange of data electronically might be critical to survival, giving employees the tools to be more productive while reducing overhead.

- Reduced delivery cycle times mean reduced lead times and lowered inventory carrying costs.

Accuracy: Accuracy in the exchange of business documents is always important. The traditional paper document exchange requires information transfer through transcription or data entry, and any such information transfer introduces errors into the process. Increases in speed are often difficult to attain because of the need to avoid transcription errors. As speed increases, so does the likelihood of introducing errors into the process. Advantages gained by increases in speed can be easily offset by the high cost of error correction. Several obvious cost savings result from increased accuracy of information transferred to suppliers and customers:

- Increased customer satisfaction

- Reduced overhead required either to detect or to reprocess erroneous documents

- Reduced costs to expedite goods or services that are late or lost

EDI Implementation

In the recent past, several large companies, including large manufacturers and retailers (such as Walmart and Target), have started mandating that their vendors be EDI-compliant before they even consider doing business with them. Thus, for many companies, EDI is a critical business enabler rather than a competitive differentiator. The traditional EDI application followed the hosted software approach. However, because many smaller suppliers either had small or nonexistent IT departments, finding EDI solutions, writing EDI maps, installing the software, carrying out integration, and then maintaining it was usually cumbersome. Coupled with the fact that EDI was itself not a revenue stream, this meant that many small businesses usually dragged their feet when it came to EDI implementation.

As a result, many of the newer vendors for EDI are moving toward the SaaS platform, and EDI is one of the key areas for SaaS adoption. In a SaaS model, the company that wants to implement the solution simply contacts the EDI provider and signs up for the service. Because the EDI provider supports thousands of large trading partners, it often already supports the required trading partner specification.

Some Well-Known EDI Providers

- Covalent
- DiCentral
- First B2B
- Perceptant
- Redtail Solutions
- WeSupply

Transportation Management System (TMS)

Traditionally, the movement of freight involved a substantial amount of paperwork, time, and involvement. Suppliers needed to keep track of every single shipment made to each downstream channel partner, payments received and invoices tendered, customs and duties (if applicable), individual freight of all kinds (FAK) breakdowns, and more. As you can imagine, shipping departments often had a hard time keeping things straight, and this often led to missed shipments, misplaced invoices, or payments not tendered or received on time. All of that led to unhappy channel partners. A *transportation management system (TMS)* helps solve these problems by streamlining several of the cumbersome activities involved in managing freight.

A TMS is a versatile software system that controls and manages various activities within the transportation channel. It understands which goods are to be shipped and received. The purpose of the TMS is to manage the process around the shipment of freight, helping the user select the right carrier across all modes, rate the movement, tender the load, print the shipping documents, track the load, bill the correct party for the freight, audit carrier invoices, and pay the freight bill from the carrier. Furthermore, the system captures and communicates relevant data on orders, shipments, rates, contracts, vehicles, shipping lanes and routes, and more. This also ensures easy access to tracking and tracing of activities in the transportation channel. Thus, a TMS is a system that helps a transportation professional make the right decisions about freight. Conceptually, a TMS is typically "positioned between" an enterprise resource planning (ERP) system (discussed in the next section) and a warehouse management system (WMS) (see Figure 6-2). In addition, some companies choose to give their suppliers access to certain parts of their TMS suite, to ease the process of booking and tracking shipments.

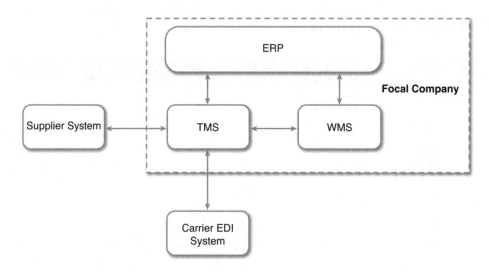

Figure 6-2 Sample IT configuration showing TMS interplay with other systems.

Benefits and Applications of TMS

Figure 6-3 shows an example of a typical TMS screen. The benefits a TMS provides are usually reduced labor cost and increased efficiency because of the core modules that are part of the software. The major modules of a typical TMS include modules that give users control over the activities involved in freight movement, such as rate shopping/rating, load tendering/carrier selection, routing and optimization, shipment tracking, shipment consolidation, payments and invoicing, reporting and scorecarding, and auditing. We discuss some of these next:

- **Rate shopping and load tendering**—Rating is one of the fundamental aspects of a good TMS. Part of what makes freight management so cumbersome without a TMS is the wide variation in carrier contract terms—especially when it comes to accessorial charges (such as lift gate fees and fuel surcharges) or FAK consolidation. Quite likely, a single company has relationships with many freight carriers, with each having its own method of charging for accessorials and specific lane treatments. In addition, certain carriers might be eligible to operate only on certain routes or negotiated lanes. It is critical that the freight manager determine which carriers are eligible to move the said freight and tender the freight accordingly. Thus, for a company/shipping manager to make accurate decisions, a TMS must have the ability to create and maintain several carriers, rate tables, accessorials, and contracts in the application. Figure 6-4 provides an example of rate shopping in a TMS.

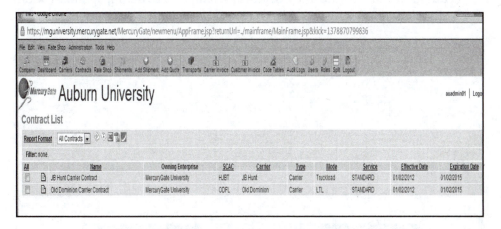

Figure 6-3   A typical TMS solution.

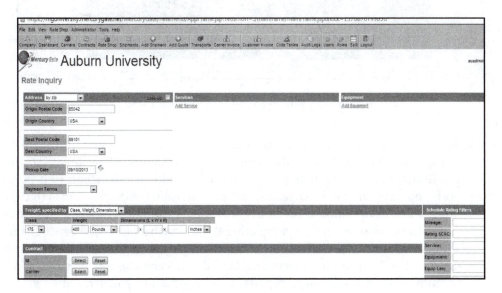

Figure 6-4   Rate shopping in a TMS.

- **Routing and optimization**—Proper planning of delivery routes has a major impact on timely order fulfillment, customer satisfaction, and long-term firm success. Thus, efficient routing and scheduling is a crucial capability TMS users seek. Good TMS systems use sophisticated mathematical algorithms and optimization routines to evaluate possible combinations in which routes could be run in the most cost- or time-efficient way possible. Some TMS solutions provide this as an integrated solution within the core TMS product; others include this as an "add-on" feature. Figure 6-5 provides an example of a routing optimizer within a TMS solution.

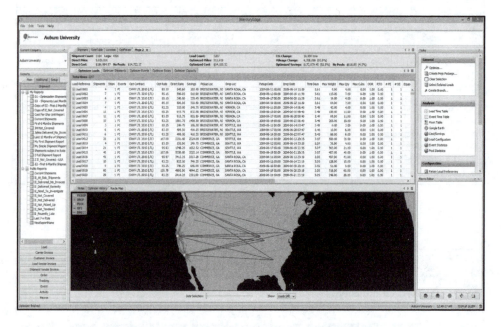

Figure 6-5   Routing application within TMS.

- **Shipment tracking**—Maintaining visibility of freight as it moves through the transportation channel is a critical part of transportation management. The in-transit status of freight can be monitored using a TMS in conjunction with global positioning system (GPS) navigation tools. In addition, TMS systems can retrieve order status information through Serial Shipping Container Codes (SSCC, discussed in the next section). Among other things, the benefit of this functionality within a TMS is to provide information to all parties in the transaction about delivery details (including potential delays, if applicable), and to ensure smooth functioning of business. Most important, linking channel partners within a TMS (see Figure 6-2) allows all the partners to view shipment status information without having to send around tracking numbers for individual orders.

■ **Payments and invoicing**—A TMS helps companies reduce manual entry of freight bills (carrier invoices), thus speeding up the process of paying and getting paid, and also eliminating errors. A common practice of TMS users is to receive freight bills electronically via EDI-equivalent messages (see the previous section on EDI). This invoice is passed directly from the carrier system to the TMS and requires no keying of data. Alternatively, if a carrier cannot pass a freight bill via EDI for some reason, often a TMS has a web portal that allows the carrier to enter the freight bill (see Figure 6-6).

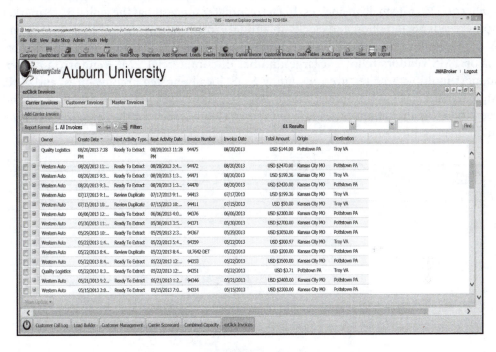

Figure 6-6   TMS screen showing invoices.

■ **Reporting, scorecarding, and auditing**—Most TMS solutions allow users to access two kinds of reports: operational and analytical. *Operational reports* allow managers to streamline freight movement during shipping; *analytical reports* provide managers with the ability to make postshipment evaluations of carrier performance, customer service, and cost. Both types of reports provide freight managers with key information with respect to future negotiations with freight carriers.

TMS Implementation

Originally, a TMS was a licensed application that followed the locally hosted or ASP architecture. This often required a significant hardware expense, in addition to the cost of ongoing system maintenance and upgrades, thus putting TMS solutions beyond the reach of many small businesses. In some cases, this is still the scenario, but the SaaS model has overwhelmingly become the "go-to" format for most commercial TMS providers. In such applications, the user logs on over the Internet to a server that the TMS provider maintains. The application resides on that server and is maintained by the software provider. The user pays for the system on a subscription or transaction basis.

The move to this type of service model has substantially enhanced the appeal and accessibility of TMS for small businesses. In fact, estimates suggest that the global TMS market now exceeds $650 million and clocks an annual growth rate of around 10 to 11 percent. A large chunk of this growth can certainly be attributed to the SaaS model that most major TMS vendors follow. A large number of commercial TMS vendors, however, support multiple architectures, including SaaS, locally hosted architectures, and ASP. MercuryGate TMS, for example, allows users to choose their preferred architecture and even change it after adoption. Figure 6-7 shows a login screen for the MercuryGate TMS in the SaaS architecture.

Figure 6-7   MercuryGate TMS SaaS login screen.

As such, the TMS marketplace sees two distinct kinds of service providers. The first are the service providers that specialize in standalone TMS solutions (such as MercuryGate TMS). Although these service providers are typically smaller, they tend to provide highly specialized solutions for TMS. The other type of service provider is large ERP vendors (such as JDA and SAP), which provide add-on TMS solutions as part of their larger suite of ERP solutions. Each of these models has its own advantages, and users should carefully study which solution best suits their needs before procurement.

Some Well-Known TMS Providers

- AdvantageTMS
- IBM Nistevo
- JDA
- LeanLogistics
- MercuryGate TMS
- TMW Systems

Routing and Scheduling (R&S) Systems

Because of the vast numbers of trucks dispatched each day, synchronization becomes critical if companies are going to provide excellent customer service. Routing and scheduling (R&S) systems help many companies ensure that orders will be delivered to the right place at the right time.

What Are R&S Systems?

Probably few technological innovations in the field of transportation have come as long a way in the past 10 to 15 years as vehicle R&S systems have. The reason for this is simple: In the late 1990s, cellular technology was still a novelty and available only to relatively limited businesses because of its high cost. In addition, software-based map databases were only beginning to be developed, and GPS technology was in its infancy and, therefore, rather expensive. Although routing software was sometimes available, data such as real-time traffic and weather conditions were hard to come by (if available). Often the mismatch between the dispatcher's available information and the driver's onground information on factors such as traffic and road maintenance conditions was large enough that a mismatch would arise between the routes drivers thought best and what the system thought best—and the driver was often more precise. This made it hard for dispatchers and traffic managers to know with any degree of certainty where their fleets were and how they were going about their routes. Finally, given that freight volumes fluctuated, planning vehicle routing efficiently was almost impossible.

Modern routing and scheduling systems allow companies, especially shippers and distributors, to efficiently manage their transportation network by intelligently allocating vehicles on lanes in such a way as to optimize cost while satisfying delivery constraints and enhancing customer service levels. Thus, R&S systems offer the promise of comprehensive "route optimization" in an automated manner by helping build that ideal mix of *orders*, *stop sequencing*, and *scheduling*, together with the shortest, most cost-efficient driving route to execute it, that will both maximize productivity for the fleet assets and maintain or improve service performance for your customers. To do this, such systems use several technologies, including real-time dynamic map displays, routing algorithms, vehicle and driver monitoring systems, and two-way communication systems. Figure 6-8 shows the workings' of a modern R&S system.

Benefits and Applications of R&S Systems

Compared to traditional dispatching methods, correctly selected and implemented R&S systems usually result in savings of about 10 to 25 percent in terms of trucks, drivers, and hours, and around 5 to 15 percent reductions in total distribution costs. Of course, the exact extent of these savings is a case-to-case issue and depends on several factors, such as the type of R&S system implemented, the extent of autonomy provided to the system, the actual nature of optimization routine adopted by the system (which is usually different in each system and is often a closely guarded secret), practical considerations, the original aim of the implementation (for example, saving costs versus improving service levels). Although there is some variability in the R&S offerings of various vendors, many of them offer some basic functional activities:

- **Functional Activity (F.A.) 1 (Pre-Dispatch)**—A key functionality that R&S systems provide is geocoding addresses and calculating routing. Stated differently, R&S systems help in locating the latitude and longitude of sites by matching the address against data contained in a digital map database; then they determine the best paths through street networks between sets of sites. As a result, they are able to provide the most efficient delivery plans for transportation companies. They do this by solving vehicle routing problems using proprietary routing algorithms that allocate an assignment of stops to routes and terminals, sequence stops, and route vehicles between pairs of stops. In addition, most R&S systems can display the results of such optimized routes in both graphical and tabular forms so that dispatchers can communicate daily route plans to drivers, loaders, and other personnel.

F.A .I: Pre-Dispatch - System provides scheduling, routing, and stop sequencing to vehicles.

F.A .2: Post-Dispatch - GPS monitors progress, real time traffic, time compliance, etc. System receives GPS input and updates vehicles in real time.

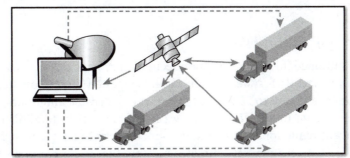

F.A .3: Post-Delivery - System receives trip feedback, driver performance and compliance data, etc. and updates database.

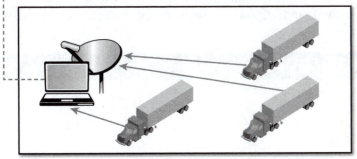

Figure 6-8 Workings of an R&S system.

- **Functional Activity (F.A.) 2 (Post-Dispatch)**—Many R&S systems are able to monitor factors such as real-time traffic conditions and driver compliance with regulations and also communicate with drivers in real time. Thus, in a way, R&S systems often give companies an "eye in the sky" as far as freight and its transport is concerned. A key element of visibility that such systems often provide is the ability to monitor vehicles remotely: For example, if something happens to a truck, from an accident to a simple flat tire, the dispatcher knows the information immediately and can quickly arrange for services to help.

- **Functional Activity (F.A.) 3 (Post-Delivery)**—Onboard computers connected with modern R&S systems often capture key information such as how a vehicle is driven, idle time, hard braking, open doors, and more. This can be used in device training programs as appropriate. For example, getting drivers to eliminate engine idling is traditionally a challenge for fleet managers, given that idling strains the engine and is prohibited in some locales. For instance, in New York City, vehicles that idle for more than 3 minutes face up to a $2,000 fine. Data that R&S systems capture has been useful in developing training programs for drivers to reduce such behavior.

R&S System Implementation

As we have mentioned, few transportation technologies have developed as much in the past 10 to 15 years as R&S systems. It has been suggested that automated route plans help companies average 10 to 25 percent fewer trucks, drivers, and hours. Additionally, companies are known to realize 5 to 15 percent reductions in total distribution costs or 8 to 20 percent reductions in miles and hours for service fleets through R&S system implementation. Current market estimates indicate that the price of an R&S solution depends on factors such as the size of the fleet and the extent of functionality desired. As such, whereas traditional R&S implementations were on hosted systems, the SaaS model seems to be catching up in this arena, with several of the newer companies preferring this approach.

Some Well-Known R&S Providers
■ DNA Evolutions
■ ESRI
■ IBM ILOG
■ MJC2
■ Optrak
■ Route Solutions
■ Telogis

Automatic Identification

To save time and increase data accuracy, many companies are moving to automatic identification applications to increase inventory visibility.

What Is Automatic Identification?

Automatic identification and data capture (AIDC) is a method of automatically identifying objects, collecting data about them, and entering that data directly into computer systems (with no human interference). Transportation technologies that are usually considered a part of AIDC include bar codes and RFID.

Bar Codes

The ubiquitous bar code is probably one of the most common technologies used in transportation management and is also one of the last things one thinks of when considering the term *high-tech*. However, a fair amount of technology goes into this rather mundane (and sometimes boring) element of the supply chain.

The original use of bar codes was to identify railcars. As the railcar rolled past a track-side scanner, it was identified and, inferentially, its destination and cargo were read. Over time, however, the bar code has been used for several other functions, including point-of-sale (POS) data capture (through the UPC/EAN/GTIN[1]), internal inventory tracking, and data capture during transportation (SSCC). Of these, the Serial Shipping Container Code (SSCC) is most relevant from a transportation standpoint, so the bulk of the discussion focuses on this topic. First, however, you must understand some of the science behind the bar code.

The most common form of a bar code is the linear bar code, wherein the data is coded as a binary code (1s and 0s) through a series of lines and spaces. The lines and spaces are of varying thicknesses and are printed in different combinations. To be scanned, the code must be accurately printed and must have adequate contrast between the bars and spaces (which is why a bar code is typically in black and white). Scanners employ various technologies to "read" codes. The two most common are lasers and cameras. Scanners can be fixed position, as with most supermarket checkout scanners, or hand-held devices, often used in taking inventories.

Coding conventions in bar codes can be of various kinds. At the retail level, the most common coding convention is the Universal Product Code (UPC), which is the coding convention used for labeling consumer products in many countries, including the United States, Canada, the United Kingdom, Australia, and New Zealand. In its most common form, the UPC consists of 12 numerical digits that are uniquely assigned to each item type. The first six to nine digits of a UPC are referred to as the *company prefix* and are assigned by a nonprofit organization (GS1). This sequence of digits uniquely identifies a company and remains constant on all its products. The next set of digits is called the *product number*. Product numbers uniquely identify individual stock-keeping units (SKUs). Unlike the GS1 company prefix, product numbers are assigned by each company and do not need to follow any set convention. The last character is called the *check digit*. Using some form of check digit generator, this digit is calculated using a mathematical calculation based on the first 11 digits of the UPC code.

Note that the UPC is not the only form of bar code available. For internal operations (nonconsumer items), especially internal inventory counting applications, companies often use other types of codes, such as Code 39. Another common coding format that finds extensive use in transportation is Code 128. The key difference between the UPC and Code 128 is that the latter was developed to accommodate letters along with numbers and can thus support alphanumeric information. It finds extensive use in the transportation industry because it can be used to encode shipping labels and mailing addresses. (For example, USPS delivery confirmation stickers are printed using Code 128.) Most kinds of information can be printed on a Code 128 bar code (see Figure 6-9).

Figure 6-9   Sample Code 128 bar code.

The Serial Shipping Container Code (SSCC): A Special Tool in Transportation Management

The SSCC is a data coding and communication standard designed to provide a standard code and symbology system that all parties (including manufacturers, transporters, distributors, and retailers) can use to track and trace shipments. The SSCC runs on the Code 128 format and was designed to support as wide a range of applications within the distribution system as possible. When coupled with shipment information provided in advance by means of EDI, the SSCC supports applications such as shipping/receiving, inventory updating, sorting, purchase order reconciliation, and shipment tracking.

In its most common form, the SSCC is a standard coding system designed to identify and label shipping containers. For the purposes of the SSCC, a container is defined as "the smallest physical unit which is not permanently attached to another unit at any point in the distribution process, and which therefore will be handled as a separate unit by the sender or recipient of goods." The beauty of the SSCC is that several different types of information can be encoded with appropriate prefixes (called *application identifiers*). For example, appropriate application identifiers on an SSCC shipping label can identify information such as a shipment's EAN/UCC article number, important variable characteristics such as the number of items in the shipment, special handling instructions, expiration dates, and more by simply scanning an SSCC-compatible bar code. The SSCC is particularly suited to identifying customer-specific product mixes, enabling better tracking of merchandise that is packed differently from one order to another, or where products are picked and packed to meet individual orders, and still have a need to be identified.

The value of the SSCC really becomes apparent when it is coupled with applications such as EDI and TMS systems in the transportation channel. Figure 6-10 illustrates this best. Note that, in the figure, solid lines represent physical movement of goods, whereas broken lines represent the virtual movement of information. The overall process can be explained as a series of steps:

Step 1: Orders are triggered at the manufacturer/vendor/upstream channel partner level, based on various customer requirements.

Step 2: Cases are prepared according to orders. Each case has its own SSCC.

Step 3: Cases are assembled into pallets. Each pallet gets a unique SSCC.

Step 4: Pallets are physically loaded onto trucks, railcars, or other vehicles while the SSCC information is virtually transmitted to the TMS.

Step 5: The bill of lading is created.

Step 6: The order information, including the SSCC, is transmitted to the customer by way of EDI in the form of an advance shipment notification.

Step 7: The customer/downstream channel partner receives the SSCC via EDI and uses it to efficiently and quickly process the shipped goods upon receipt.

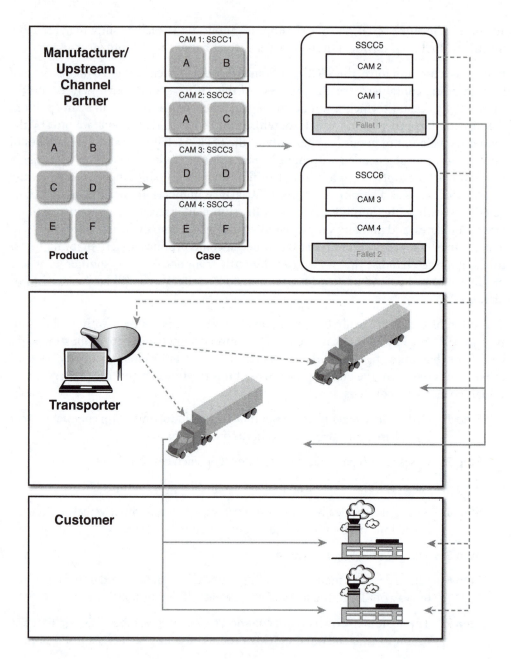

Figure 6-10   The value of the SSCC in the transportation network.

Radio Frequency Identification (RFID)

RFID is an automatic identification method that supports storing and remotely accessing data by way of specialized tags. It is important to note the distinction between the two primary types of RFID technology, active and passive tags. An RFID tag is called an *active tag* when it is equipped with a battery that can be used as a partial or complete source of power for the tag's circuitry and antenna. On the other hand, a *passive tag* does not contain a battery; its power is supplied by the reader. When a passive RFID tag encounters radio waves from the reader, the coiled antenna within the tag forms a magnetic field. The tag draws power from it, energizing the circuits in the tag. In general, active RFID tags have a greater read-distance than do passive tags. Active tags are also considerably larger than passive ones. Figure 6-11 shows a sample passive RFID tag. You likely have used RFID tags without even realizing it—for example, if you have ever paid your traffic toll using a "pass" on your car windshield, you have almost certainly used an RFID tag. Readers installed at key checkpoints (tollbooths) check for compliance and deduct the appropriate total from your account.

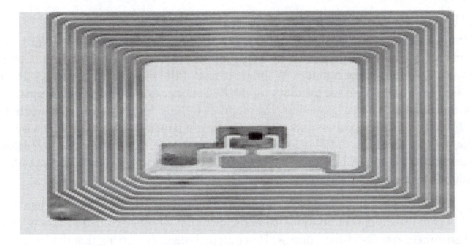

Figure 6-11   Sample RFID tag.

RFID readers can usually read only a small range of frequencies. For this reason, few different frequencies are going to be used significantly to simplify the system. Three popular frequencies are currently available on the market. Each has its own niche market that it caters to. The most robust frequency is 125 KHz; it can be read through metal, water, and practically any other surface. These chips are the most expensive, generally costing between $2 and $10 dollars, and they are typically used only on larger, more expensive items. The most common example of the 125 KHz chip is in the Mobil SpeedPass. Its range is about 4 to 6 feet in a car driving at 100 mph. A less expensive chip is the 13.56

MHz chip, which costs about 50 cents. It can transmit its signal through water but not metal and has a read range of only about 3 feet. The most common frequency in use today is 915 MHz or UHF (Ultra-High Frequency). UHF can read up to 20 feet in open air; however, it cannot penetrate water or metal.

How RFID Works

The basic working of an RFID system is simple and can be illustrated through a three-step process (see Figure 6-12). An RFID-equipped system essentially consists of three parts: an RFID tag, an RFID scanner, and a database. The RFID tag holds key information about the product in a specialized format called the Electronic Product Code (EPC) in binary form. The EPC is an extension of the basic UPC bar code that is carried by most retail items in retail stores, except that it has an added ability to store individual item information. When an RFID scanner is switched on, it generates an invisible "balloon" of electromagnetic energy. Any tags that fall within this balloon receive this energy and get "charged up." Consequently, they begin reflecting the energy in the form of data stored in them. The scanner receives this binary data and enters it into the database.

RFID Applications in the Supply Chain

Within North America, a significant portion of the rush toward RFID comes from the mandates major retailers issue to their suppliers to become Generation 2 (Gen 2)-compliant. Among these retailers, Walmart is one of the most vocal advocates driving the push toward RFID usage by adopting the EPC. Other examples include Target, Best Buy, and Staples. Walmart issued its first mandate in June 2003, wherein it mandated its top 100 suppliers to tag pallets and cases beginning in January 2005; all suppliers were to follow suit by 2006. The initial mandates were quite stringent, but it has been repeatedly suggested that Walmart has retreated at least a little in terms of its original mandated deadlines, ostensibly because of lack of buy-in at the supplier level.

It is easy to see why retailers more strongly champion the push toward RFID than other members in the supply chain: Research has shown that RFID-equipped products have a replenishment rate faster than non-RFID-equipped ones, suggesting that the benefits of the replenishment are greatest at the store level. For example, in a pilot study, Walmart stores incorporating RFID-enabled goods reported a total savings of more than $1.7 billion over similar ones that did not incorporate RFID.

For the most part, manufacturers have been slow to adopt this technology and are meeting retailer mandates based on a "slap and ship" operation, essentially adding a step to order fulfillment operations. One of the often-cited and primary benefits of RFID is that it aids in stockout reduction; 70 percent of the time, responsibility (and blame) for stockouts rest with the retailer. Although stockout reduction benefits both the manufacturer and the retailer, it seems that manufacturers are making the larger investment to improve customers' operations. Moreover, a substantial portion of the information generated by tagging rests with the end retailer, and although attempts have been made to share this information, issues such as consumer privacy have arisen, making data sharing a difficult proposition for trade partners. Finding the return on investment (ROI) for RFID has proven challenging for manufacturers and retailers alike, and it represents a significant limiting factor to widespread adoption of the technology. In general, RFID adoption varies on which stage of the supply chain you work in. According to recent research, RFID adoption rates look somewhat like Table 6-3. Note that the percentages do not add up to 100 percent because the same business process can support multiple dimensions.

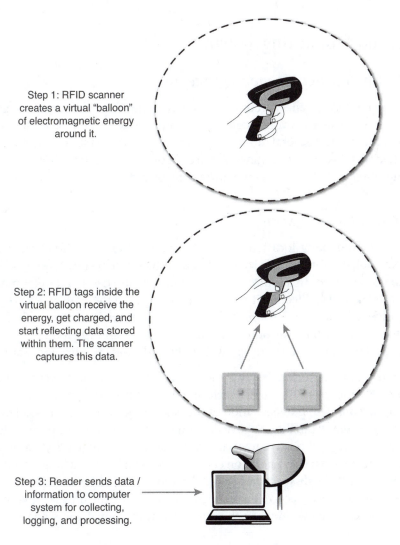

Step 1: RFID scanner creates a virtual "balloon" of electromagnetic energy around it.

Step 2: RFID tags inside the virtual balloon receive the energy, get charged, and start reflecting data stored within them. The scanner captures this data.

Step 3: Reader sends data / information to computer system for collecting, logging, and processing.

Figure 6-12 Workings of an RFID system.

Table 6-3   Frequency of References by SCM Dimension

Dimension	Number of Cases	Percent
Outbound logistics	143	57%
Inbound logistics	78	31%
Internal operations	35	14%
Returns dimension	57	23%
Others (including loss prevention)	13	5%

Control and Monitoring Systems

What Are Control and Monitoring Systems?

Control and monitoring systems represent modern technological methods of gathering data and, in some cases, performing commands and control over a vehicle, fleet, or cargo. Although in this context several different types of monitoring systems can be visualized, we restrict our discussion to overviews of two categories or monitoring systems: location monitoring systems (GPS), and temperature control and monitoring systems.

Location Monitoring Systems

Currently, two competing location monitoring systems exist in the marketplace: the U.S.-backed Global Positioning System (GPS) and the Russian Global Navigation Satellite System (GLONASS). In addition, others are in various stages of development (such as the Europe-backed Galileo Positioning System). Among these, the GPS system is by far the most widely used one, so most of our discussion revolves around it. Note that the basic technology behind the other systems is similar; if and when they become more popular, the science discussed here will still apply, with some minor modifications.

The GPS refers to a network of 31 operational satellites that orbit the earth at a height of about 12,500 miles from the earth's surface. The satellites orbit the earth at a speed of about 2.4 mph, completing one rotation of the earth about every 12 hours. This means that, on any given day, a satellite is above the same spot twice. More important, however, the satellites are arranged relative to each other in such a way that at least four satellites are visible in the sky from every point on earth at any given instant. These satellites transmit their location through specialized digital radio waves, also called *pseudo-random code*. GPS-enabled devices receive these signals and calculate the time lag involved (typically in nanoseconds) between when the satellite sent the signal and when the receiver received it. Given that radio waves are electromagnetic energy and travel at 186,000 miles per second (mps), the time lag between sending the signal and receiving it allows a GPS-enabled device to calculate the exact distance between itself and the satellite. By calculating such a distance between itself and all the visible satellites in the sky (at least four at any instant, as we mentioned earlier), a receiver can precisely pinpoint its location at that time instance. (Note that GPS receivers can typically accurately estimate their location to within about 65 feet of the real location using this approach.) This approach is known as *trilateration*. We give a brief example of trilateration in the following example—note that, to simplify the concept, we illustrate it in a 2D space. In reality, 3D trilateration works similarly.

How GPS Works

Suppose you are parachuted into a totally unknown place, maybe somewhere in the middle of a remote rural location. You run into the local gas station, buy a pack of gum, and ask the clerk where you are exactly. The only answer she is able to give you is, "You are 120 miles from Lexington, Kentucky." Although this is a useful bit of information, it still does not solve your problem. You could be anywhere on a circle with a radius of exactly 120 miles of Lexington, Kentucky, in any direction. The possibilities are endless (see Figure 6-13a). Just as you begin to wonder what to do next, the person behind you at the checkout counter says, "I know that you are exactly 85 miles from Columbus, Ohio." This second piece of information helps you pinpoint your location a little better, because there can be only two locations that are 120 miles from Lexington, Kentucky, *and* 85 miles from Columbus, Ohio (see Figure 6-13b). Now suppose that the store manager comes in and says, "You are 100 miles from Indianapolis, Indiana" (see Figure 6-13c). If you had these three pieces of information and a map of the United States, you would be able to deduce that you had parachuted into Middletown, Ohio. This is how trilateration in a GPS system works, except that it is carried out in a 3D space rather than a 2D one.

GPS Applications in Transportation

Apart from the obvious applications of GPS, including vehicle routing and real-time traffic monitoring, several other GPS applications allow transporters to manage their freight more efficiently. For example, *route adherence monitoring* is a special application of asset tracking that involves GPS. Route adherence monitoring (also called *geofencing*), uses sophisticated algorithms along with real-time location information collected via GPS to analyze and display location data, enabling commercial dispatchers and, conceivably, law enforcement officials to quickly address exceptions such as route deviations, entry to restricted areas, and developing schedule failures. Similarly, GPS technologies allow remote monitoring of drivers' adherence to such issues as compliance with speeding regulations and hours-of-service (HOS) rules.

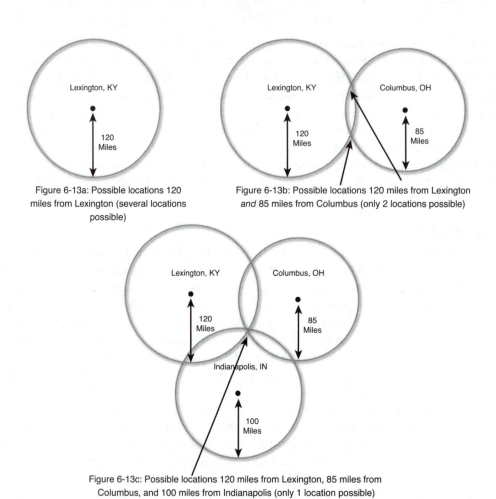

Figure 6-13a: Possible locations 120 miles from Lexington (several locations possible)

Figure 6-13b: Possible locations 120 miles from Lexington *and* 85 miles from Columbus (only 2 locations possible)

Figure 6-13c: Possible locations 120 miles from Lexington, 85 miles from Columbus, and 100 miles from Indianapolis (only 1 location possible)

Figure 6-13   How GPS works.

Temperature Control and Monitoring Systems

The transport of perishable cargo requires special thought, equipment, and care. A common agent of natural decomposition is heat, which can break down compounds to their natural state, thereby degrading them. Refrigeration throughout the transportation channel is often used to either slow down or eliminate this process of decomposition. Such a temperature-controlled supply chain is called a *cold supply chain* (cold chain). It can be understood as the transportation channel that involves the movement of temperature-sensitive items along a supply chain through thermal and refrigerated packaging methods, thereby creating a temperature-regulated environment all through the channel. Cold supply chains have several technological elements, including temperature-controlled warehouses, specialized packaging material, reefer vessels, and temperature-monitoring sensors. A discussion on temperature-controlled warehouses is beyond the scope of this chapter and will likely be covered in most standard warehousing texts. However, we cover the other elements of the cold chain in this section.

Cold Chain Packaging Technologies

Packaging technologies in temperature-controlled supply chains involve one of two types: passive shippers or active shippers. *Passive shippers* can be understood as packages that maintain a temperature-controlled environment inside an insulated enclosure, using a finite amount of preconditioned coolant in the form of chilled or frozen gel packs, phase-change materials, dry ice, or others. Passive shippers are "rated" based on the amount of time that they can hold the payload at the said temperature. Typical ratings include 24, 48, 72, and 96 hours. *Active shippers*, on the other hand, use electricity or some other fuel source to maintain a temperature-controlled environment inside an insulated enclosure under thermostatic regulation. Thus, the key difference between active and passive shippers is that whereas active shippers have some technology available to proactively cool them, passive shippers typically have no such ability. Typically, active shippers are larger in size and more expensive than passive shippers. As a result, active shippers are typically pallet-sized or larger and are useful for large, bulk cargo. Passive shippers, on the other hand, are typically smaller and lighter and are useful for smaller shipments.

Cold Chain Temperature-Monitoring Technologies

Temperature monitoring is a key element of cold supply chains. Often if the freight has been found to have violated specified temperature ranges, the recipient can reject the entire shipment. Some of the most common monitoring technologies are chemical tracer-based, RFID-based, and universal serial bus (USB)-based.

Chemical tracer-based temperature-monitoring systems are the oldest, and possibly cheapest, systems for monitoring temperature integrity and compliance throughout the transportation network. Such systems are often little more than tags treated with specialized chemicals so that they change color or show certain visible signs when they are exposed to certain temperatures. The visual sign is typically irreversible, indicating that after the tags have been "exposed," they cannot revert to their original look and color. Such tags and systems are useful in identifying *whether* temperature violations have occurred somewhere during the transportation process. However, they are less effective in identifying *when* such violations might have occurred. In addition, they are not very useful in identifying whether multiple violations have occurred. (If multiple violations do occur, usually only the first one gets recorded, because after recording the first violation, the tag is "spent.") As a result, use of such tags is decreasing.

RFID-based temperature monitors are typically small, credit card-size, semiactive tags that are preprogrammed to "fall asleep" and "wake up" at predetermined time intervals. For example, such a temperature sensor can be installed on a shipment of temperature-sensitive goods and be preprogrammed to wake up and take the temperature of the carton/pallet/container every two hours. The tag can then report the temperature to a real-time data-collection device through an RFID reader and go back to sleep for the next two hours. In essence, then, the tag can ensure temperature compliance visibility in a much more detailed and granular manner than the first type of tag. With the widespread use of RFID technology, such tags are finding substantial use in transportation.

USB-based systems gather temperature compliance data with a USB device that is connected to a port in the temperature-monitoring device (usually a digital thermometer). Upon delivery, it is removed from the device and connected to a computer's USB port. The results can then be emailed to the shipper immediately. Often such USB-based systems can work in conjunction with those based on RFID.

Summary

This chapter has highlighted several different technologies that play a role in the transportation of freight. Moreover, we have looked at the differences between technology architectures (locally hosted, ASP, SaaS) and how these relate to various transportation-related technologies (such as EDI, TMS, R&S, RFID, and control and monitoring). The field of technology is evolving rapidly, and firms are discovering new ways to leverage technological resources to drive value. The field of transportation management is no exception to this phenomenon. We therefore expect that the field will continue to evolve and that new innovations will continue to drive value in transportation management.

Key takeaways from this chapter include:

- Technology can help companies avoid the sting of the bullwhip effect.
- Technology implementation can be of three types: locally hosted, remotely hosted, or SaaS.
- EDI helps support the electronic exchange of standardized documents in electronic format directly between channel partners.
- A TMS is a specialized software tool that supports various activities within the transportation network, including rating the movement, tendering the load, printing the shipping documents, tracking the load, billing the correct party for the freight, auditing carrier invoices, and paying the freight bill from the carrier.

- R&S systems allow companies, especially shippers and distributors, to efficiently manage their transportation network by intelligently allocating vehicles on lanes to optimize cost while satisfying delivery constraints.

- Automatic identification and data capture (AIDC) methods automatically identify objects, collect data about them, and enter that data directly into computer systems. These include bar codes and RFID.

- Control and monitoring systems represent modern technological methods of gathering data and, in some cases, performing commands and control over a vehicle, fleet, or cargo. These include location monitoring and conditioning (temperature) monitoring.

Endnote

1. Universal Product Code/European Article number/Global Trade Item Number

For Further Reading

Chopra, S., and M. Sodhi (2007), "Looking for the Bang from the RFID Buck," *Supply Chain Management Review* 11(4):34-41.

Clients First Business Solutions (2011), "Cloud, SaaS and Hosted...What's the Difference?" www.erpsoftwareblog.com/2011/05/cloud-saas-and-hosted-whats-the-difference/. Accessed 24 September 2013.

Coyle, J., J. Langley, B. Gibson, R. Novack, and E. Bardi, "Supply Chain Management—A Logistics Perspective," *Cengage Publishing, 8th ed.* (New York: Cengage, 2008).

Hugos, M., *Essentials of Supply Chain Management,* 3d ed. (Hoboken, NJ: Wiley, 2011).

Farrell, J. (2013), "GPS Made Simple," VIGIL, Inc.

Kumar, S., *Connective Technologies in the Supply Chain,* 1st ed. (Boca Raton, FL: Auerbach Publications, 2007).

Rao, S. S., and T. J. Goldsby (2007), "Radio Frequency Identification in Supply Chains: Looking to Process Improvement as a Source of Financial Return," *Proceedings of the CSCMP Educator's Conference,* Council of Supply Chain Management Professionals: Lombard, IL.

Sharma, V., *Information Technology Law and Practice: Law & Emerging Technology Cyber Law & E-Commerce,* 3rd revised ed. (New Delhi, India: Universal Law Publishing Co. Ltd., 2011).

Stroh, M., *A Practical Guide to Transportation & Logistics*, 3rd ed. (Dumont, NJ: Logistics Network, Inc., 2006).

Treleven, M. D., C. A. Watts, and P. T. Hogan, "Communication Along the Supply Chain: A Survey of Manufacturers' Investment and Usage Plans for Information Technologies," *Mid-American Journal of Business* 2000;15:53–62.

Watts, C., V. Mabert, and N. Hartman, "Supply Chain Bolt-ons: Investment and Usage by Manufacturers," *International Journal of Operations & Production Management* 2008;28(12):1219–1243.

7

TRANSPORTATION'S ROLE IN LOGISTICS AND SUPPLY CHAIN STRATEGY

Transportation can factor into logistics and supply chain strategy in numerous ways. Succinctly, when transportation capabilities are lacking, the role of transportation is diffused through longer lead times for customer orders or more inventory held. When transportation is fast, safe, and reliable, however, it can serve as an important basis of competition through reduced and consistent lead times, undamaged order delivery, and reduced costs in other areas of the business, including inventory. This chapter reviews different logistics and supply chain strategies in which transportation factors significantly, including the pursuit of lean logistics; vendor-managed inventory (VMI); collaborative planning, forecasting, and replenishment (CPFR); and multivendor consolidation, among others.

Lean Logistics

Lean manufacturing has been a topic of great interest in business during the past three decades. The term *lean* was first applied to Toyota in a widely read study of the automotive industry conducted by the Massachusetts Institute of Technology in the 1980s. The study found that, by employing a production method known as the Toyota Production System (TPS), Toyota was able to produce more cars of higher quality with fewer resources (people, facilities, material, and time). The company was said to, therefore, be "lean" because it was free of the wastes that consume resources but fail to generate value for customers.[1] Since that "discovery," lean manufacturing has enjoyed a revolution in converting companies from planning-based operations that pursued economies of scale in production to flexible operations that seek to produce only what's needed when needed on a *just-in-time (JIT)* basis. In fact, an estimated 90 percent of U.S. manufacturers are employing lean as a method of continuous improvement in their operations.[2]

Manufacturing on a JIT basis carries significant implications for transportation and logistics. When inventory is regarded as a form of waste, companies seek to procure in small quantities from suppliers on a more frequent basis. Similarly, the lean manufacturer seeks to ship to customers in volumes that meet the immediate needs of customers instead of supplies that might last weeks or months into the future. Transportation, therefore, must adapt from shipping in large volumes infrequently to shipping in smaller volumes on a higher frequency. Furthermore, it becomes essential that transportation be performed at very high levels of reliability. Delays in supply deliveries can impact the manufacturer's ability to produce on a JIT manner and, in turn, its ability to deliver to customers on time. Manufacturers in the automotive industry have instilled the practice of penalizing suppliers for *every minute* that a delivery is late, even forcing suppliers to buy cars that are incomplete due to the disruption caused by late deliveries.

This shift in manufacturing philosophy can be reflected in a change of transportation modes that the supply chain might seek to employ. Instead of employing modes that support large volumes (for economy of scale), such as ocean shipping or rail, shippers resort to truck or even air transport to provide speed and reliability needed to support the JIT operations. Shippers might devise *crossdocks* (see Figure 7-1) to maintain transportation economies when using truck transportation to support higher frequencies and smaller lot sizes. A crossdock functions much like a hub airport for passenger travel. A hub airport sits quietly at the beginning of the day. Then a wave of inbound planes arrives, with passengers coming from various origins. The passengers deplane and sort themselves out, boarding planes for a vast number of destinations. Note that the planes the passengers board are the same planes that brought the collection of passengers together, although most passengers will board a different plane than the one that brought them to the hub. After the sort occurs, the planes take off for their respective destinations. Then a second wave of planes descends on the hub for another sortation. This process repeats itself over the course of the day until several sortations are achieved and the hub closes at the end of the day with all passengers cleared of the facility.

The keystone of a hub airport, and of crossdock facilities for freight, is that vehicles are full upon arriving and departing the hub. This supports transportation economies. Clearly, the challenge is for passengers (and freight) to not have to travel too far out of route to participate in the crossdock activity. We, as passengers, sometimes find it frustrating to have to travel north to a hub to reach a southern destination.

Crossdocks and hub airport operations also require a great deal of coordination in order to support customer needs at the lowest possible cost. Some describe the inflow and outflow of crossdocks as a ballet in light of the choreography required to make them work most efficiently. Consider, for instance, how frustrated one can be when missing a connecting flight due to a late arrival at a hub airport. Similarly, freight can be "frustrated" by missing departures on outbound trucks.

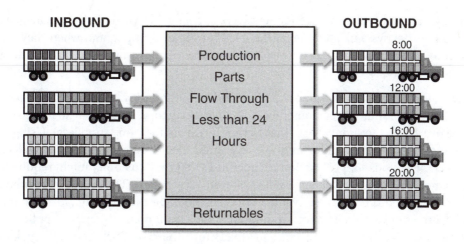

INBOUND

OUTBOUND

Production
Parts
Flow Through
Less than 24
Hours

Returnables

8:00

12:00

16:00

20:00

Figure 7-1   The crossdock operation.

To support lean logistics, some companies devise *milk runs*. Milk runs refer to circular routes that collect or drop off freight at multiple points along the way, to make efficient use of a truck. Milk run routes might be run several times over the course of a day or week, depending on the shipping volumes and required frequency necessary for serving the customer. The circular routes are a departure from the conventional one-way routes in which suppliers independently ship to the customer location. Milk runs instead require intense coordination among the suppliers to accommodate the efficient flow. As indicated previously, a delay at one location can delay delivery to the ultimate customer. Therefore, collections and drop-offs must be planned and executed precisely to support JIT operations. The more stops designed in a milk run, the more risk of delays escalates because each location offers potential for delay. Furthermore, the shipment quantity and frequency of milk runs must be adjusted from time to time to accommodate increases and decreases in customer demand. When demand decreases, the frequency of collection can decrease and/or the lot size collected can decrease. When the frequency decreases, however, the shipper must ensure that it is not holding inventory for extended periods of time.

The circular nature of milk runs supports the use of *returnable shipping containers*. Not only do trucks collect freight at supplier locations, but they also deliver empty containers to the respective suppliers for subsequent use. Lean companies, therefore, use the containers as a form of *kanban*, a visual signal of consumption that initiates replenishment. When a container is emptied, it is returned to the supplier, who then replenishes the container with goods for the next delivery. This aspect of lean logistics requires a very high level of precision, too, because containers that become lost, damaged, or delivered to the wrong location can impair the supplier's ability to replenish goods for the customer. In this way, returnable shipping containers serve as a valuable form of visual management for the logistics system. They signal the need for replenishment, and when they fail to return to the right supplier in a timely manner, they signal a defect in the planning or execution processes.

Note that milk runs can be run on the inbound or outbound side of a crossdock to support the flows of multiple customer locations. Many customers are likely to have considerable overlap in the supply base (that is, they buy from the same suppliers). Using milk runs, the inbound logistics for multiple customers can be accommodated by collecting goods at the various supply locations in quantities that each customer might require for a short amount of time (perhaps a few days or even a few hours). The collection of inbound milk runs (called *subroutes*) convenes at the crossdock for the sortation, forming truckload volumes of vast assortments of different goods from many different suppliers. The premise here, again, is one of full truckloads into the crossdock and full truckloads outbound from it. What each customer receives, then, is an assortment of goods from several different suppliers on one truck. This helps to provide higher levels of in-stock performance on these items (because the assortment arrives more frequently than if the individual store ordered less frequently) and lower costs of operation and administration because the store must receive only one truck for the assortment instead of having each manufacturer ship separately.

As these examples illustrate, lean principles are finding ready application in transportation and logistics. In some cases, "lean logistics" means logistical support for lean manufacturing, by providing JIT deliveries. In other instances, firms are employing lean principles in logistics to identify and eliminate the various forms of waste that might be found in the focal business and the supply chain. Often wastes are found in the transportation system that can be eliminated to reduce costs and, potentially, improve the service offered to customers. The next sections refer to related forms of innovation in transportation that seek to improve service and reduce costs.

Shared Transportation Resources

The pursuit of lean logistics has inspired shippers and service providers to seek innovative ways of reducing waste in transportation and logistics. It is troubling to realize that every shipper has a degree of underutilized transportation capacity. Not only are half-empty trucks moving down the road, but completely empty trucks are moving as a means to reposition equipment, either deadheading back to a terminal location or seeking out the next load. In total, this amounts to an enormous amount of wasted transportation capacity across the millions of shippers within a nation or around the world. The philosophy associated with lean logistics, described previously, is empowering companies to seek ways to eliminate these wastes and to more effectively use the available transportation assets. Some competitors are going a step further by devising ways to reduce waste and reduce environmental impact through *horizontal collaboration*. In these examples, companies bound for common customer locations share transportation capacity.

The trend of horizontal collaboration has proven most progressive among grocery manufacturers in Europe. Companies such as Colgate-Palmolive, GlaxoSmithKline, Kellogg's, Kimberly-Clark, Mars, Nestlé, Procter & Gamble, and Unilever are sharing transportation capacity to reduce duplication in journeys, empty miles, and poor load factors. Together these efforts yield improvements in service for customers while reducing fuel consumption, operating expenses, and carbon emissions.[3] Horizontal collaborations garner antitrust concerns in some countries, however, for fear of creating unfair advantages.

In the United States, companies work through third parties to devise innovative collaboration programs. Third-party logistics providers (3PLs) have devised these *multivendor consolidation* systems to support grocery retailers. The retailers find that such consolidation is an effective way to reduce out-of-stocks, shipping costs, damage claims, and administrative costs. To take the concept to a higher order, some 3PLs now offer *multivendor/multiretailer consolidation* to leverage the combined purchase volumes of multiple, independent retailers into a converged inbound logistics system. Converged volumes make higher delivery frequencies economically feasible across the various locations of different retailers. As one can imagine, it also makes the planning more complex because the third party must now accommodate multiple customers who are also likely to be head-to-head competitors through a common logistics system. Smaller retailers might find such a collaborative strategy effective for competing with much larger competitors that can enjoy economies of scale on their own.

Merge-in-Transit (MIT)

A related form of collaboration to multivendor consolidation, *merge-in-transit (MIT)* is an increasingly popular concept in this era of rising shipping costs. The basic idea behind MIT is intuitive: An MIT system unites shipments from various suppliers at a merge point. Typically, such a merge point is located close to the end customer, to allow multiple suppliers to "merge" at that central location. MIT allows businesses to trim transportation expenses by consolidating their less-than-truckload (LTL) shipments with another vendor's compatible freight headed to the same customer. By consolidating freight at an intermediate *merge point*, vendors can reduce their reliance on the length of LTL travel. At the customer level, MIT provides value because it facilitates the delivery of a single, consolidated shipment rather than multiple smaller shipments. In addition, the ability to cost-effectively transport smaller quantities of goods means that vendors can be convinced more readily to ship in smaller batches, thus reducing overall inventory in the system and supporting lean manufacturing.

Figure 7-2 shows the typical MIT operation. As you can see, the MIT system depends on several key factors, including the availability of multiple vendors for a customer, regularity of shipments to the customer, order size (the most frequent order size for this system varies between 150 and 10,000 lbs), and the availability of a strategically located merge point. When these elements have been identified, an MIT operation can be created. The system begins at the individual vendor level. Instead of shipping direct from origin shipper to consignee, orders are shipped on trailers direct to regional terminals (merge points). There, the order is offloaded, segregated, and merged by consignee. Finally, the order is then reloaded (typically onto full truckloads) for delivery to the ultimate destinations. As a result, instead of receiving multiple LTL shipments, the end customer receives a reduced number of shipments.

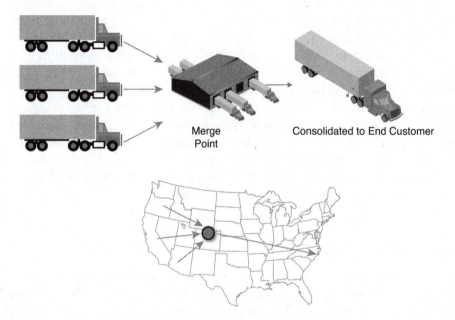

Merge Point

Consolidated to End Customer

Figure 7-2 Merge-in-transit (MIT) operation.

Although the MIT operation seems appealing, its implementation has several challenges. First, as is obvious, such system involves multiple loadings and unloadings of freight. This has the potential to increase overall freight cost if is not optimized correctly. The more times freight is "touched" in transit, the higher the cost (all else being equal). Continuing with the example, the more touch points freight has, the more likely the chance that the goods will be damaged. Finally, given that an MIT system involves consolidated deliveries, the system relies on the assumption that the vendors will fulfill the order-filling operation at the same level of efficiency and on the assumption of cooperation between vendors. For example, a delay in receiving a shipment from one vendor at the merge point would arguably delay the shipment to the end customer. In that sense, the success of the MIT system is only as strong as the weakest link in the network.

The exact opposite of MIT is a system called *pooled distribution*. In such a system, one vendor typically distributes orders to numerous destination points within a particular geographic region. Instead of shipping direct from origin shipper to consignee, orders are shipped on consolidated trailers direct to regional terminals. There the pool is offloaded, segregated, sorted by consignee, and then reloaded onto local delivery trucks for delivery to the ultimate destinations.

Vendor-Managed Inventory (VMI)

One of the most pervasive supply chain strategies of recent years is *vendor-managed inventory (VMI)*. VMI assumes many different forms, but the general premise involves the supplier managing the inventory of the customer in a supplier–customer dyad. Under a VMI arrangement, the supplier and customer agree to the terms of replenishment (such as how much inventory the customer is willing to hold and the timing and volume of replenishment). The customer then holds the supplier accountable for maintaining the inventory levels within the prescribed specifications. Such arrangements dramatically alter the typical buy–sell arrangements in business. Instead of "pushing" inventory to customers on an infrequent basis and finding motivation in selling large volumes to customers that might take months for the customer to use or consume, the seller in VMI arrangements seeks to provide only as much inventory as is prescribed in the relationship. This usually results in small, more frequent deliveries, similar to the provisions of lean logistics described previously. In fact, it is argued that VMI lends to a leaner supply chain.

One variant of VMI is *vendor-owned and -managed inventory (VOMI)*. Under this arrangement, the supplier not only manages the inventory residing at a customer location, but maintains ownership of it as well. Conventionally, this arrangement is considered a consignment arrangement because the transaction between buyer and seller does not occur until the buyer actually uses or consumes the inventory. The term *scan-based trading* is used to refer to these arrangements in retail environments, where scanning an item at the point of checkout signals a complete transaction not only between the retailer and the consumer, but also between the retailer and the vendor of the item. VOMI arrangements alter the buy–sell arrangements between customers and suppliers even more than conventional VMI, in that the inventory at a customer location resides on the financial books of the supplier. With that in mind, the supplier is motivated to maintain lean inventory levels at the customer location and to monitor those inventory levels closely.

The transportation implications of VMI and VOMI arrangements include a shift in the responsibility of planning inventory, replenishment, and load planning for the vendor (supplier) in the relationship. This can have significant implications on not only the frequency and timing of deliveries, but also the coordination. For instance, can a supplier be allowed to schedule deliveries at a customer location? Typically, the customer controls its own facilities and receiving yards. However, if a vendor is held accountable for stockouts at the customer location, the vendor might be granted more influence in how it is allowed to come and go from the facility. Furthermore, vendor representatives (including delivery personnel) are often not allowed in a customer facility. Yet it might be important for a delivery person or customer relationship manager to see the inventories residing at the customer facility, to ensure that they are ample in volume and quality. These are matters that must be resolved in comanaged inventory arrangements that are not found in conventional buy–sell relationships in which the rules of engagement have been in place for many years.

Collaborative Planning, Forecasting, and Replenishment (CPFR)

Whereas VMI focuses on replenishment, higher orders of supply chain collaboration examine opportunities for team effort in the planning and forecasting steps that precede replenishment operations. Hence, collaborative planning, forecasting, and replenishment (CPFR)[4] is a strategy for trading partners in the supply chain to work together to coordinate their marketing and logistics activity for improved service and lower costs. Under CPFR, a supplier and customer develop joint business plans, agreeing on the objectives for conducting joint business and the means by which the two companies will collaborate. The successive steps focus on getting the two firms aligned in sales forecasts, resolving differences, and acting in concert to ensure a coordinated effort. The nine-step CPFR process follows[5]:

1. Develop a front-end agreement.
2. Create the joint business plan.
3. Create the sales forecast.
4. Identify exceptions for the sales forecast.
5. Resolve/collaborate on exception items.
6. Create an order forecast.
7. Identify exceptions for the order forecast.
8. Resolve/collaborate on exception items.
9. Generate the order.

The premise of CPFR is to align the planning and operations that occur between trading partners to reduce uncertainty and the costs of doing business. Research has shown CPFR to be an effective means of reducing the *bullwhip effect*, a well-documented phenomenon that results in stockouts when a change in demand occurs and dramatic overstocks arise from poor coordination among trading partners. Many companies have found success by extending sales and operations planning (S&OP) with trading partners (suppliers and customers) through the CPFR method.[6] Each year, the Voluntary Interindustry Commerce Solutions (VICS) organization recognizes best practices in implementation.[7]

Collaborative Transportation Management (CTM)

An extension of the CPFR logic is found in *collaborative transportation management (CTM)*, another initiative of the VICS organization. Whereas CPFR focuses on the business transpiring between a supplier and a customer in the supply chain, one important party is largely left out of these planning conversations: the logistics service provider that facilitates the transportation of goods from supplier to customer. CTM seeks to remedy this exclusion by including logistics providers in the coordination of customer deliveries. CTM is defined as "a holistic process that brings together supply chain trading partners and service providers to drive inefficiencies out of the transport planning and execution process."[8] The premise is to eliminate the waste from the processes of the trading partners and the service provider in pursuit of win-win-win outcomes for the three parties.

In the final step of the CPFR process (step 9, order generation), CTM picks up from an operational sense, providing focus on the distribution and transportation of goods. However, much of the benefit in CTM is associated with involving the logistics service provider in the planning of shipments. Whereas the CPFR process devised overall sales forecasts and order forecasts, 3PLs and carriers can use these as inputs to devise a *shipment forecast*. Such planning allows the service provider to allocate equipment and labor for upcoming shipments. Without this information, dramatic changes in shipping demands are sprung on the service provider, even though the two trading partners might know of the changes well in advance. By eliminating these surprises, carriers are better able to ensure capacity to the trading partners for their mutual business and to reduce the anxiety and costs associated with emergency actions. The service provider can plan for changes and even provide input that can influence the plans of the trading partners.

Figure 7-3 (a and b) shows the details of the CTM process. The sequence of activities is labeled as strategic, tactical, and operational. At the *strategic* level, the parties to the collaboration identify how to do business together and illustrate the objectives of the collaborative effort, similar to the early steps in CPFR. At the *tactical* stage, the firms work together to devise the shipment forecast, including resolving any exceptions or differences that are perceived across the parties. Here the carrier speaks up if there is fear of insufficient capacity to meet the anticipated volumes. This might be the case during peak shipping seasons. In the absence of CTM, the trading partners might not be aware of this constraint, and carriers would have no opportunity to voice concerns. In the final (*operational*) stage of activity, the forecasts transition to actual orders and action on the parts of all three parties. The benefit of CTM in this stage is the defined business process, particularly when issues arise calling for collective action and remedies. Without a defined process, a flurry of unproductive activity can consume the three firms, leading to errors and distracting the businesses from other activities in need of attention. In the final step of the CTM process, the parties evaluate performance against established objectives, learn from mistakes, and refine processes for future engagement.

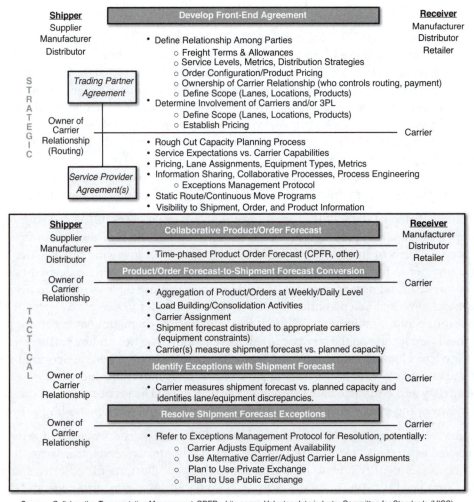

Source: *Collaborative Transportation Management*, CPFR white paper, Voluntary Interindustry Committee for Standards (VICS).

Figure 7-3a The integrated CTM business process (strategic and tactical).

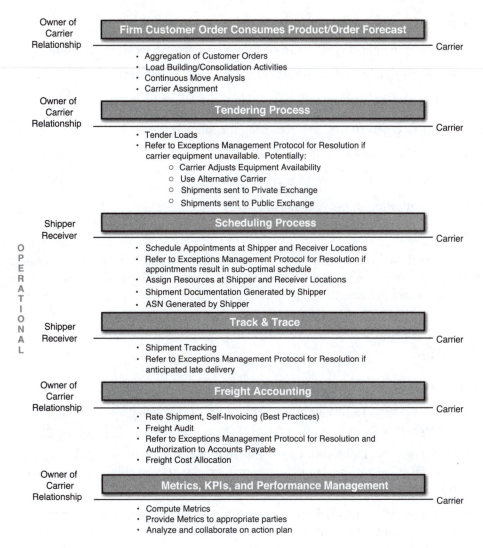

Owner of Carrier Relationship

Firm Customer Order Consumes Product/Order Forecast

Carrier

- Aggregation of Customer Orders
- Load Building/Consolidation Activities
- Continuous Move Analysis
- Carrier Assignment

Owner of Carrier Relationship

Tendering Process

Carrier

- Tender Loads
- Refer to Exceptions Management Protocol for Resolution if carrier equipment unavailable. Potentially:
  - Carrier Adjusts Equipment Availability
  - Use Alternative Carrier
  - Shipments sent to Private Exchange
  - Shipments sent to Public Exchange

Shipper Receiver

Scheduling Process

Carrier

- Schedule Appointments at Shipper and Receiver Locations
- Refer to Exceptions Management Protocol for Resolution if appointments result in sub-optimal schedule
- Assign Resources at Shipper and Receiver Locations
- Shipment Documentation Generated by Shipper
- ASN Generated by Shipper

Shipper Receiver

Track & Trace

Carrier

- Shipment Tracking
- Refer to Exceptions Management Protocol for Resolution if anticipated late delivery

Owner of Carrier Relationship

Freight Accounting

Carrier

- Rate Shipment, Self-Invoicing (Best Practices)
- Freight Audit
- Refer to Exceptions Management Protocol for Resolution and Authorization to Accounts Payable
- Freight Cost Allocation

Owner of Carrier Relationship

Metrics, KPIs, and Performance Management

Carrier

- Compute Metrics
- Provide Metrics to appropriate parties
- Analyze and collaborate on action plan

O P E R A T I O N A L

Source: *Collaborative Transportation Management*, CPFR white paper, Voluntary Interindustry Committee for Standards (VICS).

Figure 7-3b   The integrated CTM business process (operational).

The benefits accrued by participating CTM companies are impressive. Shippers and carriers report the following benefits:

- On-time service improvements by 35 percent

- Lead-time reductions of more than 75 percent (for example, average lead time for one customer was reduced from 7 days to 1.5 days)

- Inventory reductions of 50 percent

- Sales improvements of 23 percent through improved service to customers

- Premium freight cost reductions of greater than 20 percent

- Administrative cost reductions of 20 percent

Carriers, too, find benefit in CTM, reporting the following[9]:

- Deadhead mile reductions of 15 percent

- Dwell time reductions of 15 percent

- Fleet utilization improvements of 33 percent

- Driver turnover reductions of 15 percent

Results along these lines require significant coordination and time invested among the participating parties. Yet the results warrant consideration in light of companies' efforts to eliminate waste, reduce costs, and improve service.

Summary

Key takeaways from this chapter include:

- Transportation factors significantly in innovations in logistics and supply chain management.

- Lean logistics involves eliminating waste in various forms throughout the business. Effective transportation management can reduce or eliminate waste in material flows in the supply chain.

- Milk runs and crossdocks offer means for achieving consolidated shipping volumes in support of higher frequency deliveries.

- Companies are sharing transportation assets in different ways to eliminate duplicate journeys, reduce empty miles, and improve load factors and equipment utilization.

- Shippers, receivers, and service providers are collaborating in ways that change traditional business relationships and alter the ways transportation is managed in the supply chain.

Endnotes

1. James P. Womack, Daniel T. Jones, and Daniel Roos, *The Machine That Changed the World: The Story of Lean Production* (New York: Harper Perennial, 1991).

2. Aberdeen Group, *The Lean Benchmark Report: Closing the Reality Gap,* 2006.

3. EyeForTransport, "Companies who want a long-term competitive advantage are pooling supply chain assets NOW - will you be left behind?" Available at http://events.eft.com/hc-usa/index.shtml.

4. CPFR is a trademark of the Voluntary Interindustry Commerce Solutions (VICS).

5. Voluntary Interindustry Commerce Solutions (VICS), www.vics.org.

6. CPFR case studies can be found at www.vics.org/guidelines/cpfr_roadmap_case_studies/.

7. Best practice award winners can be found at www.vics.org/vics-achievement-awards-2012/.

8. *Collaborative Transportation Management*, CPFR white paper, Voluntary Interindustry Committee for Standards (VICS), 2004.

9. Ibid.

8

TRANSPORTATION AND GLOBAL SUPPLY CHAINS

International trade has been growing at a phenomenal rate ever since the end of World War II. Some of the institutions that were created at the end of the war were specifically designed to stimulate and grow trade between different nations of the world. The Bretton Woods Conference in July 1944 led to the creation of these entities:

- The International Monetary Fund (IMF), which managed the stability of exchange rates and established an international system of payment

- The World Bank, which has changed its focus from World War II reconstruction to poverty alleviation through trade and development

- The General Agreement on Tariffs and Trade (GATT), whose aim was to reduce tariffs and duties across the world to promote world trade, and to guarantee "Most Favored Nation (MFN)" status to all its members, thus preventing discrimination between locally owned firms and foreign firms

- The World Trade Organization (WTO), a successor organization to GATT created on January 1, 1995, whose goal is to "to help producers of goods and services, exporters, and importers conduct their business, while allowing governments to meet social and environmental objectives"[1]

In addition, various countries have floated trade blocs to enhance trade within specific geographic areas. Some of the prominent trade blocs are the European Union (EU), Association of South East Asian Nations (ASEAN), Gulf Cooperation Council (GCC), Southern Common Market (MERCOSUR), North American Free Trade Agreement (NAFTA), and South African Customs Union.

International trade can be divided between merchandise trade and trade-in services. Merchandise trade involves export and import of physical goods. Service trade involves the export and import of intangible products such as people skills, finance, education, legal, transcription, entertainment, tourism, and communications. As far as the transportation industry and global supply chains are concerned, merchandise trade plays a more important role, compared to the trade in services. Table 8-1 gives a global overview of world merchandise trade and the share of some of the bigger economies of the world in this trade.

Table 8-1   Total Merchandise Trade in Value and Percentages for the World and Select Economies.

	Value, mn $	Share		Annual Percentage Change		
	2011	2005	2011	2005-2011	2010	2011
Total exports f.o.b.	17,816,372	100	100	10	22	20
Total imports c.i.f.	18,057,065	100	100	9	21	19
Leading Exporters						
European Union (27)	6,038,597	40	34	7	12	17
extra-EU (27) exports	2,132,888	13	12	8	17	19
China	1,898,381	7	11	16	31	20
United States	1,480,432	9	8	9	21	16
Japan	822,564	6	5	6	33	7
Korea, Republic of	555,214	3	3	12	28	19
Russian Federation	522,013	2	3	14	32	30
Hong Kong, China	455,650	—	—	8	22	14
domestic exports	16,832	0	0	−3	−12	14
re-exports	438,818	—	—	8	23	14
Canada	452,440	4	3	4	23	17
Singapore	409,503	2	2	10	30	16
domestic exports	223,688	1	1	10	33	23
re-exports	185,590	1	1	10	28	10
Leading Importers						
European Union (27)	6,255,558	40	35	7	13	17
extra-EU (27) imports	2,349,849	14	13	8	19	17
United States	2,265,894	16	13	5	23	15
China	1,743,484	6	10	18	39	25
Japan	854,998	5	5	9	26	23
Korea, Republic of	524,413	2	3	12	32	23
Hong Kong, China	510,855	—	—	9	25	16
retained imports	130,237	1	1	10	27	16

	Value, mn $	Share		Annual Percentage Change		
	2011	2005	2011	2005–2011	2010	2011
Canada	462,635	3	3	6	22	15
India	462,633	1	3	22	36	32
Singapore	365,770	2	2	11	26	18
retained imports	180,180	1	1	11	24	27
Mexico	361,068	2	2	8	28	16

Source: WTO, International Trade Statistics (ITS) 2012 (www.wto.org/english/res_e/statis_e/its2012_e/its12_toc_e.htm).

Merchandise trade is dominated in both value and volume by the manufactured goods, followed by fuels and mining products, and finally by agricultural products. Figures 8-1a and 8-1b track the rise of value of the merchandise exports by absolute value and volume, respectively. Apart from the global recessionary years (2008–2010), there has generally been an uptick in the absolute value and volume of all three kinds of merchandise exports, which has led to increasing demands on the various modes of transportation worldwide.

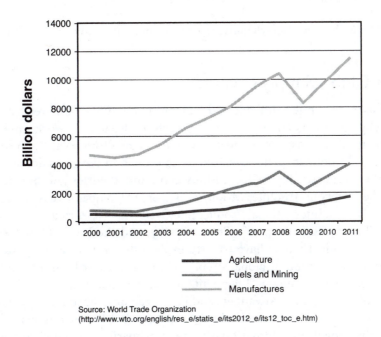

Source: World Trade Organization
(http://www.wto.org/english/res_e/statis_e/its2012_e/its12_toc_e.htm)

Figure 8-1a **World merchandise exports by value (2000–2011)...**

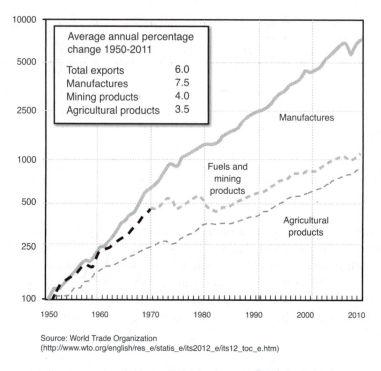

Average annual percentage
change 1950-2011

Total exports	6.0
Manufactures	7.5
Mining products	4.0
Agricultural products	3.5

Source: World Trade Organization
(http://www.wto.org/english/res_e/statis_e/its2012_e/its12_toc_e.htm)

Figure 8-1b   ...and merchandise trade by commodity volume (1950–2010).

Need for Global Supply Chains

International trade is often difficult and frustrating because of differences in culture, language, religious beliefs, traditions, morals, customs, legal framework, environmental concerns, political systems, and monetary systems. Since World War II, however, international trade has flourished and even outpaced growth in the gross domestic product (GDP) of the world (see Figure 8-2). Many economic theories and business models (including the Theory of Absolute Advantage, Theory of Comparative Advantage, Factor Endowment Theory, International Product Life Cycle Theory, Porter's Diamond Model, New Trade Theory, Ricardo-Sraffa Trade Theory, and the International Production Fragmentation Trade Theory) have tried to explain the growth in international trade that has led to the creation of global supply chains. The Ricardo-Sraffa Trade Theory and the International Production Fragmentation Trade Theory are extensions of Ricardo's Theory of Comparative Advantage and have been developed recently to overcome the limitations of the Theory of Absolute Advantage and Factor Endowment Theory. In essence, the growth in international trade is primarily attributed to comparative advantages that different nations have over one another and the ability of products to add value in different nations to keep the price of the final product as low as possible. This has enabled textile-trading firms such as Li & Fung to have different parts of the clothing supply chain in South Korea, Taiwan, Bangladesh, Thailand, and India, while catering to demand mainly in the United States and Western Europe. To enable complex global supply chains, firms such as Li & Fung rely extensively on efficient and effective modes of transportation.

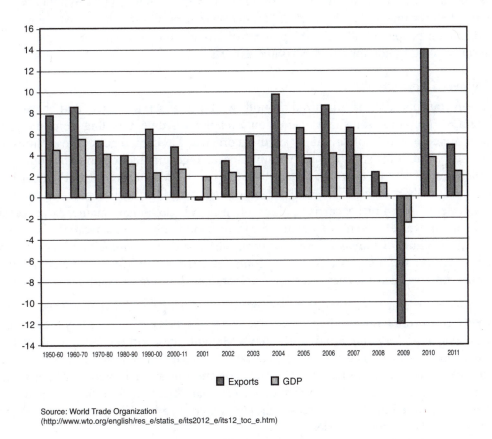

Figure 8-2 Growth of world merchandise exports and GDP (1950–2011).

International Modes of Transportation

When shipping globally, mode selection becomes critical because of high costliness, and there are many options transportation managers can choose from.

International Ocean Transportation

Of all the modes of transportation, the most important one as far as international trade is concerned is ocean transportation. Oceangoing vessels carry around 80 percent of world merchandise trade by volume and 70 percent by value. In 2011, the total volume of goods loaded worldwide was 8.7 billion tons. The world fleet of ships grew 37 percent from 2008 levels to 1.5 billion deadweight tons (dwt) in January 2012. Container traffic as measured by 20-foot equivalent units (TEUs) was 572.8 million TEUs in 2011.[2] In the United States, one in six jobs is related to marine transportation: The industry employs 2.3 million people. In addition, 95 percent of foreign trade is carried by ships, which accounts for 2 billion tons of imports and exports.[3]

Types of Service

Ocean transportation is organized around two types of service: liners and charterers (tramps). *Liners* are ships that operate on a regular schedule, traveling from one predetermined port to another. Liners are often organized as conferences, whose aim is to provide service on a specific route within a specified geographic region under uniform freight rates. They are a form of cartel whose aims are to "facilitate the orderly expansion of world sea-borne trade."[4] The bill of lading (B/L) issued by the master of the vessel is the evidence of contract of carriage. Liners often carry break-bulk/container cargo. Most B/Ls are drawn between the carrier and the shipper, and international treaties often govern the liabilities of damages. Several intermediaries, such as freight forwarders, custom house agents, and non-vessel operating common carriers (NVOCC), help facilitate the transaction between a carrier and shipper. Container ships are an example of liners.

Charterers are ships that operate under conditions of supply and demand. They do not have a fixed route, nor do they have a fixed port of call. These ships carry break-bulk and bulk cargo (dry and liquid) and often set freight rates with reference to the Baltic Exchange. Normally, a charter-party (contract) is drawn between the charterer and the ship owner as evidence of contract of transportation or affreightment. All terms and conditions and references to international treaties are mentioned in the charter-party. The B/L is issued by the charterer—examples are crude oil carriers, grain carriers, car carriers, and commodity carriers.

The most common types of charter-parties are the voyage charter, time charter, and bareboat charter/demise charter. As the name implies, a *voyage charter* refers to a contract based on chartering a ship from port of origin to port of destination. A gross form of voyage charter obliges the ship owner to take on the responsibility of paying for cargo loading, cargo discharging, and trimming (stability) of the ship. In a net form of voyage charter, the charterer pays the cargo loading, cargo unloading, and trimming charges. Some key clauses incorporated in a voyage charter are laytime (period of time to load and unload cargo from a ship without penalty) and demurrage (penalty imposed when the laytime is exceeded). Sometimes a *cesser clause* is inserted into a charter party when the charterer and the shipper on the B/L are different. The cesser clause ensures that the charterer is discharged of all the liabilities in case of any dues, and the cargo of the shipper of the B/L is held as collateral for the dues owed to the ship owner.

A *time charter* involves hiring the ship based on time and is independent of the voyage. The charterers take complete control of the ship and are responsible for all the liabilities and damages during the charter period. The ship owner normally takes an advance, to safeguard his or her interest in this specific charter-party.

A *bareboat charter*, or *demise charter*, occurs when the charter hires only the ship/vessel, without any fuel, provisions, cargo, or insurance. The charterer has complete control over the vessel and is required to pay for any incidental expenses, such as fuel, provisions, cargo, or insurance. A variation of a demise charter allows for hire-purchase in the shipping industry. This form of bareboat charter is initiated by the charterer for an extended period of time, with the intention of owning the vessel at the end of the charter time.

Types of Cargo

The types of cargo carried in oceangoing vessels are diverse and normally do not easily fit into a neat classification system. However, for the sake of simplicity, cargo is normally classified as bulk (wet or dry), break-bulk, and containerized cargo. Figure 8-3 shows the growth in different types of cargo over the years.

Bulk cargo is usually transported unpackaged. It is normally poured or dropped into the hold of the ship. Sometimes palletized or boxed cargo also is considered bulk cargo. *Break-bulk cargo* is normally noncontainerized, unitized cargo in which individual pieces are loaded on the ship.

Containerized cargos are normally of fixed lengths (TEUs or 40-foot equivalent units [FEUs]). Cargo is normally stuffed into these standard containers and then loaded into the ships. As an example, corn shipped in the holds of the ship or shipped as bags placed in holds of the ship is considered bulk cargo. The same corn packed in individual drums and shipped would be considered break-bulk cargo. If the corn were packed in TEU containers, it would be considered containerized cargo. The share of containerized cargo is increasing over the years, as Figure 8-3 shows, and the share of break-bulk cargo has been decreasing over the years. Over the years, low container prices, coupled with the financial crisis, has prompted consolidation within the container shipping industry (see Figure 8-4). A smaller number of companies tend to deploy larger container ships. Three companies, Maersk, MSC, and CMACGM, control 30 percent of the world container traffic.

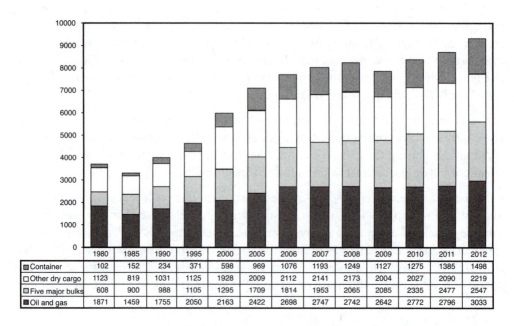

	1980	1985	1990	1995	2000	2005	2006	2007	2008	2009	2010	2011	2012
Container	102	152	234	371	598	969	1076	1193	1249	1127	1275	1385	1498
Other dry cargo	1123	819	1031	1125	1928	2009	2112	2141	2173	2004	2027	2090	2219
Five major bulks	608	900	988	1105	1295	1709	1814	1953	2065	2085	2335	2477	2547
Oil and gas	1871	1459	1755	2050	2163	2422	2698	2747	2742	2642	2772	2796	3033

**Figure 8-3   International seaborne trade, by cargo type,
selected years (millions of tons loaded).**

Source: United Nations Conference on Trade and Development (UNCTAD), Review of Maritime Transport 2012
(http://unctad.org/en/PublicationsLibrary/rmt2012_en.pdf).

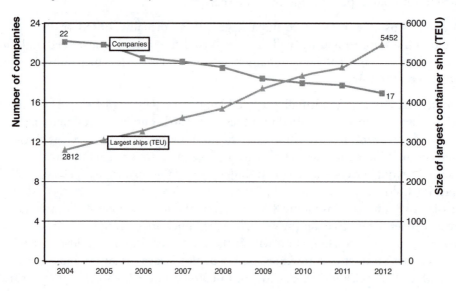

Source: Calculations by the UNCTAD secretariat, on the basis of data supplied by *Lloyd's List Intelligence.*

Figure 8-4 International seaborne industry profile.

Source: RMT 2012 (UNCTAD Review of Maritime Transport).

Types of Vessels

Vessels are classified based on either the types of cargo they carry or their size. Note that some ships fit into more than one classification system, depending on their usage and the kind of cargo they carry.

Classification based on size follows:

- **Handysize**—Ships that typically serve small and large ports, including coastal shipping. They represent the bulk of the number of ships in the world. Their capacity is between 15,000 and 35,000 dwt and can carry dry and wet cargo.

- **Handymax**—The workhorses of dry bulk cargo, with a capacity of less than 60,000 dwt.

- **Supramax**—Ships normally used in costal shipping and small ports. They often carry dry bulk cargo with a capacity of between 50,000 and 60,000 dwt.

- **Panamax**—The largest ships to transverse the Panama Canal. The highest capacity is capped at 65,000 dwt by the Panama Canal Authority. The typical size of Panamax container ships is around 5,000 TEUs. Ship length is restricted to 275 meters, with a width of 32 meters and a draught of 12.04 meters. They mainly carry commodities.

- **Aframax**—Ships named after the Average Freight Rate Assessment (AFRA) tanker rate system. They typically carry crude oil, with a capacity of between 75,000 and 115,000 dwt.

- **Suezmax**—Ships designed to navigate the Suez Canal. They can easily dock at most ports in the world and are used for containerized dry and wet cargo. They have typical capacity ranges between 120,000 and 200,000 dwt.

- **Capesize**—Ships that generally cannot go through either the Panama Canal or the Suez Canal—they must transit through either Cape Horn or the Cape of Good Hope. Capacity ranges from 80,000 to 175,000 dwt.

- **New Panamax ships**—Bigger and more efficient ships, with larger dimensions, thanks to the building of the New Panama Canal. They have the capacity to carry 15,000 TEUs. The lengths of ships through the New Panama Canal can be a maximum of 427 meters, with a width of 55 meters and a draught of 18.30 meters.

- **Post New Panamax**—Large container ships (Triple E Class) being built for Maersk, with capacities reaching 18,000 TEUs. Plans also include building container ships to hold 20,000 TEUs.

- **Very large ore carrier (VLOC)/ultra large ore carrier (ULOC)**—Iron ore carriers that travel primarily from Brazil to Europe and Asia. VLOCs are typically greater than 200,000 dwt; ULOCs are greater than 300,000 dwt. The largest ULOC ever built, the Berge Stahl in 1986, had a capacity of 365,000 dwt, a length of 343 meters, a width of 65 meters, and a draught of 25 meters.

- **Very large crude carriers (VLCC)/ultra large crude carriers (ULCC)**—VLCCs typically have a capacity of between 180,000 and 320,000 dwt. They serve the North Sea, Mediterranean, and West African Ports. ULCCs have a capacity greater than 320,000 dwt and carry crude from the Middle East to Europe, Asia, and North America. Knock Nevis, the largest ULCC ever built, had a capacity of 564,763 dwt, with a length of 458 meters, a width of 68.8 meters, and a draught of 29.8 meters.

Some ships are classified according to the unique geographical area they serve or the specialized cargo they carry:

- **Malaccamax**—These ships navigate the Strait of Malacca. They can have a maximum length of 400 meters, a width of 59 meters, and a draught of 14.5 meters.

- **Seawaymax**—Ships that can pass the locks of St. Lawrence Seaway can have a maximum length of 225.6 meters, a width of 23.80 meters, and a draught of 7.92 meters.

- **Q-max**—These ships are named after Qatar and carry liquefied natural gas (LNG) from Qatar to the rest of the world.

- **Roll-on/roll-off (RORO)**—Self-propelled vehicles, railroad cars, and livestock often are carried by specialized ships called RORO. The advantages are that the cargo does not need specialized cranes or gantries to load or unload, so labor is often cheaper. The big disadvantage is that, in most cases, the ships return to the port of origin empty because they are not capable of handling different kinds of cargo.

- **Container ships**—Approximately 60 percent of world trade by value is containerized. Container ships come in different sizes—the most common can carry 5,000 to 6,000 TEUs. The newer container ships, called the Triple E Class ships, will have the capacity to carry 18,000 TEUs.

Flags

Per international law, every ship must be registered in a country and must fly the "flag" of that specific country. Ship owners are allowed to choose which country they want to register their ships in. This has led to "open registry," with countries allowing any ship to be registered in that country. The concept of flying the flag of a country entitles the ship to be treated as an extension of the country. The naval force of the flagged country is responsible for the security of the merchant ships. In addition, the laws and regulations of the country in which the ship is registered apply on the ship. In return, the ship must pay all the taxes and dues in the country of registration. Not surprisingly, ship owners choose "flags of convenience," based on the most relaxed regulations and lowest taxation rates. Figure 8-5 gives the percentage of foreign and domestic ownership of ships registered in select countries.

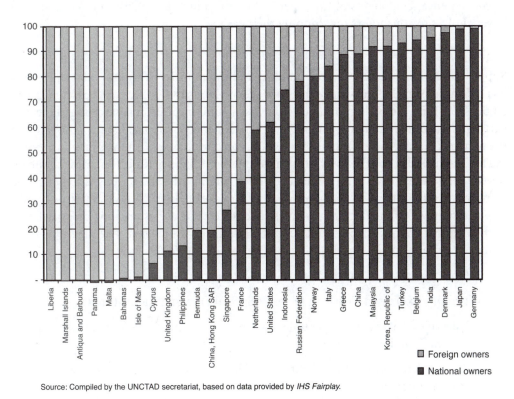

Source: Compiled by the UNCTAD secretariat, based on data provided by *IHS Fairplay*.

Figure 8-5   Foreign and national ownership of the top 30 fleets by flags of registration, 2012 (percentage share of fleet dwt).

Source: UNCTAD, Review of Maritime Transport 2012.

Cabotage

Originally, the term *cabotage* (from the French word *caboter*, meaning "to sail along the coast") was the exclusive right of a nation to navigate its inland waters. The country could decide which entities could ply the inland or coastal waterways. This term has now been expanded to cover all forms of transportation. Within the United States, the Jones

Act authorizes the movement of cargo between two U.S. ports exclusively on U.S. flagged ships. Proponents of the Jones Act and cabotage point to the fact that 500,000 jobs and $100 billion in annual economic output accrue directly to the United States. In case of war, the United States has the ability to requisition its domestic U.S. flagged fleet for the movement of troops, equipment, and provisions. The safety record of ships flagged in developed countries tends to be greater than the safety record of ships flying the flags of convenience. Opponents of the Jones Act point to increasing cost of operations and fewer choices to consumers for shipping their cargo.

Liability

Supply chains have become more complex, and the number of multimodal shipments has increased over the years since the Carriage of Goods by Sea Act was passed in 1936. At the same time, technology and newer forms of communication have changed the way cargo and documentation are generated, arranged, and transported when compared to the previous century. The United Nations attempted to change with the times by proposing the Hague Rules, the Hague-Visby Rules, and the Hamburg Rules, but most countries never ratified these documents. The latest effort by the United Nations is to adopt the Convention on Contract for the International Carriage of Goods Wholly or Partly by Sea, popularly known as the Rotterdam Rules. The Rotterdam Rules are now being debated in the member countries and could be ratified in the coming years. The salient features of the Rotterdam Rules as applied to global supply chains follow:

- One contract of carriage for multimodal transportation
- Liability expressed clearly in terms of Special Drawing Rights (SDR)
- New rules that take into account e-commerce, enhanced navigational devices, and new ways of doing business
- A clear description of the rights and obligations of the shipper, the ship owner, and the receiver

International Air Transportation

The global air network has been growing phenomenally during the last 50 years: The air network size has been basically doubling every 15 years since the 1970s. The number of passengers in 2011 rose to 3 billion, and 49.2 million tons of freight was carried in 2011. Figures 8-6a and 8-6b give the growth rate of passengers (in terms of passenger-kilometers) and cargo (in terms of freight ton-kilometer) for select years.

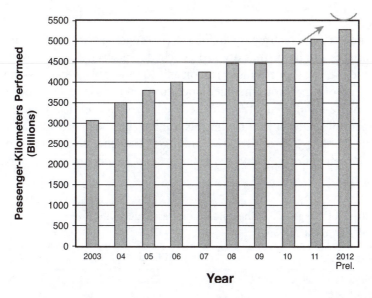

Figure 8-6a   Total scheduled traffic in passenger-kilometers performed (2003–2012).

Source: International Civil Aviation Organization (ICAO), Annual Report of the Council 2012.

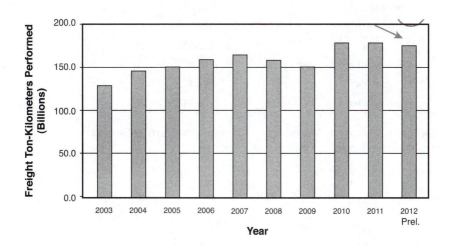

Figure 8-6b Total scheduled freight traffic (2003–2012).

Source: ICAO, Annual Report of the Council 2012.

Service

The airline industry can be broadly divided as passenger services and cargo operations. Passenger services can be further divided as scheduled services and nonscheduled services. Scheduled services are normally run on a fixed time table, regardless of the demand in a short period of time. Nonscheduled operators cater mostly to the tourism industry and private business and leisure travel. Scheduled passenger services typically accounted for 93.7 percent of the total traffic, and this number has been growing as a percentage of total traffic. Table 8-2 gives the 2012 ranking of the scheduled airlines, based on the passenger-kilometers flown. Table 8-3 gives the top 25 airports in 2012, based on the total passengers.

Table 8-2   2012 Rankings of Airlines, Based on Passenger-Kilometers (International + Domestic) Flown

Rank	Airline	Millions
1	United Airlines	288,282
2	Delta Airlines	271,567
3	American Airlines	203,336
4	Emirates	180,880
5	Lufthansa	142,512
6	Southwest Airlines	137,708
7	Air France	135,821
8	China Southern Airlines	135,021
9	British Airways	124,318
10	Qantas Airways	108,051

Source: International Air Transport Association (IATA), Scheduled Passengers – Kilometres Flown (www.iata.org/publications/Pages/wats-passenger-km.aspx).

Table 8-3 2012 Top 25 Airports, Based on Total Passengers

Rank No.	City	Airport	Passengers Embarked and Disembarked[a]			Aircraft Movements[b]		
			2012 (thousands)	2011 (thousands)	2012/2011 (%)	2012 (thousands)	2011 (thousands)	2012/2011 (%)
1	Atlanta, GA	Hartsfield-Jackson Atlanta International	95,487	92,389	3.4	930	924	0.7
2	Beijing	Beijing Capital International	81,929	78,675	4.1	557	533	4.5
3	London	Heathrow	69,983	69,391	0.9	471	476	-1.0
4	Tokyo	Haneda (Tokyo International)	66,795	62,585	6.7	391	380	3.0
5	Chicago, IL	O'Hare International	66,835	66,806	0.0	878	879	-0.1
6	Los Angeles, CA	Los Angeles International	63,688	61,862	3.0	605	604	0.3
7	Paris	Charles de Gaulle	61,612	60,971	1.1	498	514	-3.2
8	Dallas/Fort Worth, TX	Dallas-Fort Worth International	58,591	57,774	1.4	650	647	0.5
9	Jakarta	Jakarta Soekarno-Hatta International	57,773	51,533	12.1	380	346	10.1
10	Dubai	Dubai International	57,685	50,978	13.2	344	326	5.5
11	Frankfurt	Frankfurt	57,520	56,436	1.9	482	487	-1.0
12	Hong Kong	Hong Kong International	56,062	53,329	5.1	362	344	5.1
13	Denver, CO	Denver International	53,156	52,849	0.6	618	635	-2.6
14	Bangkok	Bangkok Suvarnabhumi International	53,002	47,911	10.6	317	305	3.8
15	Singapore	Changi	51,182	46,544	10.0	325	302	7.6
16	Amsterdam	Schiphol Amsterdam	51,036	49,755	2.6	438	437	0.2

			Passengers Embarked and Disembarked[a]			Aircraft Movements[b]		
Rank No.	City	Airport	2012 (thousands)	2011 (thousands)	2012/2011 (%)	2012 (thousands)	2011 (thousands)	2012/2011 (%)
17	New York, NY	John F. Kennedy International	50,819	49,198	3.3	402	409	–1.8
18	Guangzhou	Guangzhou Baiyun International	48,309	45,040	7.3	373	349	6.9
19	Madrid	Barajas	45,195	49,671	–9.0	373	429	–13.1
20	Istanbul	Ataturk International	44,999	37,395	20.3	349	302	15.6
21	Shanghai	Shanghai Pudong International	44,880	41,448	8.3	362	344	5.1
22	San Francisco, CA	San Francisco International	44,477	41,045	8.4	425	404	5.2
23	Charlotte, NC	Charlotte-Douglas International	41,228	39,044	5.6	552	540	2.3
24	Las Vegas, NV	McCarran International Las Vegas	41,668	41,480	0.5	528	532	–0.7
25	Phoenix, AZ	Sky Harbor International	40,422	40,592	–0.4	450	462	–2.6
		Total	1,404,332	1,334,699	4.4	12,061	11,909	1.3

[a] Revenue and nonrevenue air carrier passengers and passengers in direct transit; scheduled and nonscheduled services.
[b] All aircraft movements (commercial and noncommercial).

Source: ICAO, Annual Report of the Council 2012.

The cargo operations of airlines can generally be divided as scheduled air freight services and chartered air freight services. Examples of scheduled air freight services are FedEx and UPS Airlines, which operate on fixed routes on a fixed schedule. Examples of chartered air freight services are the Dreamlifter of Boeing and the Beluga of Airbus, which transport cargo when there is demand. Approximately $6.4 trillion of cargo, representing 35 percent of the total world trade by value, is flown. Table 8-4 gives the ranking of scheduled airlines, based on the freight-ton kilometers flown in 2012.

Table 8-4   2012 Rankings of Airlines, Based on Freight-Ton Kilometers (International + Domestic) Flown

Rank	Airline	Millions
1	FedEx	16,108
2	UPS Airlines	10,416
3	Emirates	9,319
4	Cathay Pacific Airways	8,433
5	Korean Air Lines	8,144
6	Lufthansa	7,175
7	Singapore Airlines	6,694
8	British Airways	4,732
9	China Airlines	4,538
10	Eva Air	4,470

Source: IATA, Scheduled Freight Ton-Kilometers.

Leases

Both passenger and cargo aircrafts can be bought outright by various operators. But in most cases, leasing companies buy the aircrafts and lease them to various operators. This helps the operators reduce capital expenditure and gives them the ability to adjust the capacity of the aircraft, depending on the demand, without much financial commitment. The only downside is in the form of higher lease fees. Leases can be of three different kinds:

- **Dry lease**—The lessor provides the lessee with only the aircraft.

- **Wet lease**—The lessor provides the lessee with the aircraft, crew, insurance, maintenance, and fuel.

- **ACMI lease**—The lessor provides the lessee with the aircraft, crew, maintenance, and insurance (ACMI).

Tariffs and Liabilities

The airline industry normally charges tariffs on the principle of charging what the market can bear. As with ocean transportation, the tariffs are normally charged on the basis of volume or weight, whichever is higher. About 240 airlines, or 84 percent of total air traffic, are represented by the International Air Traffic Association (IATA). One of the functions of IATA is to act as a clearinghouse for its member airlines and help ensure seamless travel for passengers across the world.

The liability of damages to cargo and loss of life or incapacitation to passengers has been restricted over time. Currently, the Montreal Protocol of 2009 is in force and limits liability in terms of special drawing rights (SDRs).

Open Skies

Cabotage and the rules under IATA and the International Civil Aviation Organization (ICAO) give nations the power to restrict the number of domestic and foreign airlines flying into their territory. Most international flights are negotiated between countries on a bilateral basis. However, in 1992 the United States and Norway implemented the open skies policy, to remove a restriction on the number of flights, routes, and capacity. All EU members put the same policy into effect in 2007. The only restriction now is the number of available landing slots at some of the key airports. This open skies policy has led to more airlines flying into the United States and Europe, and also has resulted in alliances being developed, such as between British Airways and American Airlines, and between Air France /KLM and Delta Airlines.

Intermodal Transportation

The primary modes of international shipment are oceangoing vessels and air. But land-based transportation such as trucking and rail are important modes of transportation in their own right. Trucking and rail often provide domestic transportation, but they also provide last-mile connectivity for international transportation. For land-locked countries, trucks and rails are the only means of connecting to global supply chains.

The critical issues for a shipper to consider before hiring trucks to move goods in any country include these:

- Weight restriction on the truck
- Weight restriction in the geographical area where it is operated
- Hours of operation governed by local laws

- Quality of infrastructure (roads, bridges, congestion, docks)
- Size of trucks and trailers that can be used

Issues that shippers normally consider before sending goods through rail include these:

- Ownership of railroads (private versus public, and the availability of subsidies to move goods)
- Infrastructure available (including the gauge of tracks, electrification of lines, maintenance of the system, availability of loading and unloading docks and platforms, connectivity to the road network, and speed at which trains can travel)
- Relationship between passenger traffic and merchandise traffic, and the priority of one over the other

Within the United States, the number of air carriers has been decreasing steadily since 1990 (see Table 8-5) because of bankruptcy, mergers and acquisitions (M&As), and reduced capacity in terms of seats offered and routes flown. The number of Class I railroads that carry the bulk of the traffic in terms of both volume and value has decreased due to M&As. An exception to the decreasing trend has been trucking (motor carriers), which has shown an increase in the number of carriers plying the nation's roadways. This could be because of the fragmented nature of the business—ownership of trucks continues to be fragmented, with a preference for owner-operators. The inland and coastal vessel operators have seen a steady decrease because of the shift in traffic away from waterways to trucks and rail. These vessel operators also are more prone to vagaries of nature, such as droughts and floods. The pipeline operators are more insulated from market forces because they specialize in moving goods such as oil, gas, and viscous chemicals in bulk at very low costs.

Table 8-5   Number of Carriers in Different Modes of Transportation for Select Years

	1990	1995	2000	2005	2010
Air carriers[a]	70	96	91	85	77
Major air carriers	14	11	15	17	21
Other air carriers	56	85	76	68	56
Railroads	530	541	560	560	565
Class I railroads	14	11	8	7	7
Other railroads	516	530	552	553	558
Interstate motor carriers[b]	216,000	346,000	560,393	679,744	739,421

	1990	1995	2000	2005	2010
Marine vessel operators[c]	U	1,381	1,114	733	603
Pipeline operators[d]	2,198	2,367	2,157	(R) 2,329	(P) 2,219
Hazardous liquids	171	197	220	(R) 308	(P) 350
Natural gas transmission	866	975	844	975	(P) 981
Natural gas distribution	1,382	1,444	1,363	(R) 1,388	(P) 1,241

KEY: P = preliminary data; R = revised data; U = data unavailable.

Source: U.S. Department of Transportation (DOT), Research and Innovative Technology Administration (RITA), Bureau of Transportation Statistics (BTS), Table 1-2: Number of Air Carriers, Railroads, Interstate Motor Carriers, Marine Vessel Operators, and Pipeline Operators (www.rita.dot.gov/bts/sites/rita.dot.gov.bts/files/publications/national_transportation_statistics/html/table_01_02.html).

[a] Carrier groups are categorized based on their annual operating revenues as major, national, large regional, and medium regional. The thresholds were last adjusted July 1, 1999, and the threshold for *Major air carriers* is currently $1 billion. The *Other air carriers* category contains all national, large regional, and medium regional air carriers. Beginning in 2003, regional air carriers are not required to report financial data which may result in underreporting of *Other air carriers* in this table.

[b] 1990-2005 figures are for the fiscal year, October through September. 2006-2009 figures are snapshots dated Dec. 22, 2006; Dec. 21, 2007; Dec. 19, 2008, and Dec. 18, 2009. 2010 figure is the U.S. DOT number of active interstate motor carriers as of the end of December 2010. The numbers of *Interstate motor carriers* are based on "active" U.S. DOT numbers. The Federal Motor Carrier Safety Administration deletes motor carriers from the Motor Carrier Management Information System (MCMIS) when they receive an official notice of a change in status. However, some companies may go out of business without de-activating their U.S. DOT number. As a result, inactive carriers may be included in the MCMIS.

[c] The printed source materials do not contain totals for the number of operators, and data files from which the figures can be determined are not available prior to 1993.

[d] There is some overlap among the operators for the pipeline modes. Therefore, the total number of *Pipeline operators* is lower than the sum for the three pipeline modes.

Intermodal transportation involves shipment of the same merchandise through more than one mode of transportation with a single rate. The Rotterdam Rules specifically provide and enhance documentation, security, and liability terms for intermodal (multimodal) transportation. Some of the features of intermodal transportation follow:

- The cargo is not handled multiple times; the container/pallet gets handled multiple times.

- A single multimodal/intermodal B/L covers all modes of transport to provide "door-to-door" delivery.

- Technology such as electronic data interchange (EDI)/Internet enables coordination among transportation agencies and various arms of governments (such as Customs and port authorities).

- Containerization of cargo is in standard units of TEUs/FEUs, for compatibility across various modes of transportation across the world. This ensures the integrity of goods inside the container.

- Cargo from different shippers is consolidated in a container, to bring down overall shipping costs.

- The benefits of different modes of transportation, in terms of costs and service levels, are utilized to optimize shipping time.

- Economies of scale from speed and flexibility have reduced costs over bulk shipments by more than 20 times.

- Warehousing and security are enhanced because the shipping unit or container is opened only at the destination.

Table 8-6 summarizes the factors that have led to increased intermodal transportation.

Table 8-6   Causes for Intermodal Transportation

Factor	Cause	Consequence
Technology	Containerization and IT	Modal and intermodal innovations; tracking shipments and managing fleets
Capital investments	Returns on investments	Highs costs and long amortization; improved utilization to lessen capital costs
Alliances and M&A	Deregulation	Easier contractual agreements; joint ownership
Commodity chains	Globalization	Coordination of transportation and production (integrated demand)
Networks	Consolidation and interconnection	Multiplying effect

Source: Jean-Paul Rodrigue, *Intermodal Transportation and Integrated Transport Systems: Spaces, Networks and Flows* (unpublished working paper, 2006), p. 10.

Even within the United States, trucks and rail play a prominent role in international trade, along with other modes of transportation. Primarily, truck and rail are used in Canada extensively and, in a limited way, in Mexico. Trucks from both Mexico and the United States are allowed only to a certain point within the borders because of the limitations in the NAFTA agreement. Trucks and rail are free to move within the United States and Canada, as far as international trade is concerned. Table 8-7 gives the volume and value breakup of the imported and exported merchandise goods.

Table 8-7   U.S. International Merchandise Trade by Transportation Mode, 2010 (Billions of U.S. Dollars)

	Billions of U.S. Dollars		
Mode	Total Trade	Exports	Imports
Water	1,434	455	979
Air	837	393	444
Truck	557	285	272
Rail	131	46	85
Pipeline	63	5	58
Other and unknown	167	94	73
Total, all modes	**3,190**	**1,278**	**1,912**
	Millions of Metric Tons		
Mode	Total Trade	Exports	Imports
Water	1,302	521	782
Air	<1	<1	<1
Truck	170	92	78
Rail	122	54	67
Pipeline	96	8	88
Other and unknown	8	5	2
Total, all modes	**1,698**	**681**	**1,017**

Source: U.S. DOT, Federal Highway Administration (FHWA), Freight Management and Operations, Freight Facts and Figures 2011 (www.ops.fhwa.dot.gov/freight/freight_analysis/nat_freight_stats/docs/11factsfigures/figure2_2.htm).

Land Bridges

One consequence of the growth of intermodal transportation has been the exploration of new ways of shipping goods to lower overall costs and also decrease shipping time. One such innovation is the concept of *land bridges*. As an example, the continent of North America is an obstacle for oceangoing vessels between the ports on the east coast of Asia and the ports on the west coast of Europe. To overcome this obstacle, intermodal firms are now using the United States and Canada as land bridges, whereby goods in containers are brought from, say, Shanghai to Seattle by sea and then sent first from Seattle to New York by rail and then from New York to Rotterdam by sea. In this case, the United States acts as a land bridge between the Pacific Ocean and the Atlantic Ocean. A big issue cropping up in land bridges is the availability of surplus capacity of infrastructure and manpower to handle increased volume of shipments not bound for the domestic market.

Enablers and Key Players in Transporting Goods to Global Supply Chains

Freight forwarders: Sending merchandise across global supply chains is complex because of differences in customs, duties, tariff and nontariff barriers, regulations, documentation, and coordination between government and transportation agencies. Specialized agencies have evolved to deal with the complexities. Freight forwarders are intermediaries who specialize in booking space for the cargo, creating the necessary documentation, and calculating the total cost of shipment.

Custom brokers and custom house agents (CHAs): These agencies often specialize in clearing goods for export and import with the customs of the country. They are well versed with the documentation needs and the regulatory environment of the law of the land.

Non-vessel operating common carriers (NVOCC): The ship owners of the air freighters, liners, and charterers rarely interact with full-container-load (FCL)/less-than-container-load (LCL) shippers, unless the shippers ship substantial volume in the ship owners' fleet. The ship owners often sell space in bulk to entities called NVOCCs, who then resell the space to individual customers. The NVOCCs in the United States are regulated by the Federal Maritime Commission. In most cases, the functions of freight forwarder, custom broker, and NVOCC might reside in the same organization, to provide seamless services to the shipper.

3PLs: A third-party logistics provider (3PL) supplies functional services, compared to 1PL (shippers) and 2PL (asset-based carriers). The functions a 3PL provides its clients could take the form of pick and pack, basic inventory management, warehousing, and distribution. The latest innovation of 3PL is to venture into the area of non-asset-based solutions such as consultation on packaging and transport, freight contract negotiations, auditing and tracking of freight, financial settlement with different parts of the supply chain, and dispute resolution with customers.

4PLs: The fourth-party logistics provider (4PL) role is still evolving, and its definition is still being contested. Some believe 4PL to be a totally non-asset-based software solutions company that provides a common software and hardware platform to all its clients, to enable the 4PLs to leverage savings from consolidation of loads, shipments, transportation, and supply chain efficiency from different shippers. Another school of thought defines 4PLs as firms that provide all the functions of a 3PL but that also have elements of non-asset-based solutions to promote commerce.

Incoterms

Incoterms are rules created by the International Chamber of Commerce, headquartered in Paris, France. The current version, called Incoterms 2010 (shown in Figure 8-7a, b), following the descriptions), came into effect January 1, 2011. Incoterms are terms of trade that are often incorporated in contracts (domestic and international) to clarify issues of risk, costs, and ownership of property. Contracts using Incoterms 2010 carry the same meaning for each three-letter abbreviation anywhere in the world. The current version has 11 Incoterms. A B/L, airway bill (AWB), or multimodal B/L is the transportation document that prominently features the Incoterm in addition to the contract itself.

The 11 Incoterms and a brief description for each follow. *Carrier* mentioned in the Incoterms can be substituted for *first carrier,* in case of a multimodal transportation.

EXW: Ex-Works (EXW *place*)

The seller is responsible for making the goods available at the seller's premises.

Costs: The buyer bears all costs, from seller's premises to destination, including transportation.

Risk and ownership of goods: Risk passes to the buyer as soon as goods are made available to the buyer. Ownership of goods transfers from seller to buyer along with the risk.

Mode of transportation used: Air, rail, road, multimodal, containerized, sea, inland waterway.

FCA: Free Carrier (FCA *place*)

The seller clears the goods for export and hands them over to the carrier and place named by the buyer.

Costs: The buyer bears all costs, from the time the goods are handed over to the carrier, at the place named by the buyer.

Risk and ownership of goods: Risk passes to the buyer as soon as goods are handed over to the carrier at the place named by the buyer. Ownership of goods transfers from seller to buyer along with the risk.

Mode of transportation used: Air, rail, road, multimodal, containerized, sea, inland waterway.

FAS: Free Alongside Ship (FAS *loading port*)

The seller clears the goods for export and places them alongside the ship at the named port.

Costs: The buyer bears all costs from the time the goods are placed alongside the ship, including loading of the cargo.

Risk and ownership of goods: Risk passes to the buyer as soon as goods are placed alongside the ship at the port named by the buyer. Ownership of goods transfers from seller to buyer along with the risk.

Mode of transportation used: Sea, inland waterway.

FOB: Free On Board (FOB *loading port*)

The seller clears the goods for export and places them on board the ship at the named port.

Costs: The buyer bears all costs from the time the goods are placed on board the ship. The seller pays for loading of the cargo.

Risk and ownership of goods: Risk passes to the buyer as soon as goods are placed on board the ship at the port named by the buyer. Ownership of goods transfers from seller to buyer along with the risk.

Mode of transportation used: Sea, inland waterway.

CFR: Cost and Freight (CFR *destination port*)

The seller clears the goods for export and pays all charges, including freight, up to the destination port mentioned by the buyer.

Costs: The buyer bears all costs from the time the goods reach the destination port. The seller pays for loading and freight of the cargo up to the destination port.

Risk and ownership of goods: Risk passes to the buyer as soon as goods are placed on board the ship at the port of origin. Ownership of goods transfers from seller to buyer along with the risk.

Mode of transportation used: Sea, inland waterway.

CIF: Cost, Insurance, and Freight (CIF *destination port*)

The seller clears the goods for export and pays all charges, including freight, up to the destination port mentioned by the buyer. The seller buys insurance on behalf of the buyer.

Costs: The buyer bears all costs from the time the goods reach the destination port. The seller pays for loading, freight, and insurance of the cargo up to the destination port.

Risk and ownership of goods: Risk passes to the buyer as soon as goods are placed on board the ship at the port of origin. Ownership of goods transfers from seller to buyer along with the risk.

Mode of transportation used: Sea, inland waterway.

CPT: Carriage Paid To (CPT *place of destination*)

The seller clears the goods for export and pays all charges, including freight, up to the place of destination mentioned by the buyer.

Costs: The buyer bears all costs from the time the goods reach the place of destination. The seller pays for loading and freight of the cargo up to the place of destination.

Risk and ownership of goods: Risk passes to the buyer as soon as goods are handed over to the carrier at the place of origin. Ownership of goods transfers from seller to buyer along with the risk.

Mode of transportation used: Air, rail, road, multimodal, containerized, sea, inland waterway.

CIP: Carriage and Insurance Paid to (CIP *place of destination*)

The seller clears the goods for export and pays all charges, including freight, up to the place of destination mentioned by the buyer. The seller buys insurance on behalf of the buyer.

Costs: The buyer bears all costs from the time the goods reach the place of destination. The seller pays for loading, freight, and insurance of the cargo up to the place of destination.

Risk and ownership of goods: Risk passes to the buyer as soon as goods are handed over to the carrier at the place of origin. Ownership of goods transfers from seller to buyer along with the risk.

Mode of transportation used: Air, rail, road, multimodal, containerized, sea, inland waterway.

DAP: Delivery at Place (DAP *place of destination*)

The seller is discharged of all responsibilities when the goods are placed at the disposal of the buyer on the mode of transportation on which it has arrived and is ready for unloading at the place of destination.

Costs: The seller bears all costs until the goods are ready to be unloaded at the place of destination from a mode of transportation (including insurance, if applicable). The buyer pays for unloading goods and every other charge subsequent to unloading.

Risk and ownership of goods: The seller bears all risks up to the place of destination and the point of unloading. Ownership of goods transfers from seller to buyer along with the risk.

Mode of transportation used: Air, rail, road, multimodal, containerized, sea, inland waterway.

DAT: Delivered at Terminal (DAT *place of destination*)

The seller is discharged of all responsibilities when the goods are placed at the disposal of the buyer at the buyer's terminal of choice after unloading from the mode of transportation. "Terminal" includes a place, whether covered or not, such as a quay, warehouse, container yard, or road, rail, or air cargo terminal.

Costs: The seller bears all costs up to the unloading of goods at the buyer's terminal of choice. The buyer pays for all costs subsequent to unloading the goods.

Risk and ownership of goods: The seller bears all risks up to the place of destination and including the unloading of goods. Ownership of goods transfers from seller to buyer along with the risk.

Mode of transportation used: Air, rail, road, multimodal, containerized, sea, inland waterway.

DDP: Delivery Duty Paid (DDP *place of destination*)

The seller is discharged of all responsibilities when the goods are placed at the disposal of the buyer at the place of destination requested by the buyer. All export and import Customs clearance are the responsibility of the seller.

Costs: The seller bears all costs until the goods are placed at the disposal of the buyer. Costs include loading, unloading, tariffs, Customs duties, and any other incidental charges required to be paid to deliver the goods to the place of destination of the buyer.

Risk and ownership of goods: The seller bears all risks up to the place of destination of the buyer until the buyer takes charge of the goods. Ownership of goods transfers from seller to buyer along with the risk.

Mode of transportation used: Air, rail, road, multimodal, containerized, sea, inland waterway.Source: ICAO, Annual Report of the Council 2012.

INCOTERMS® 2010:
The International Chamber of Commerce

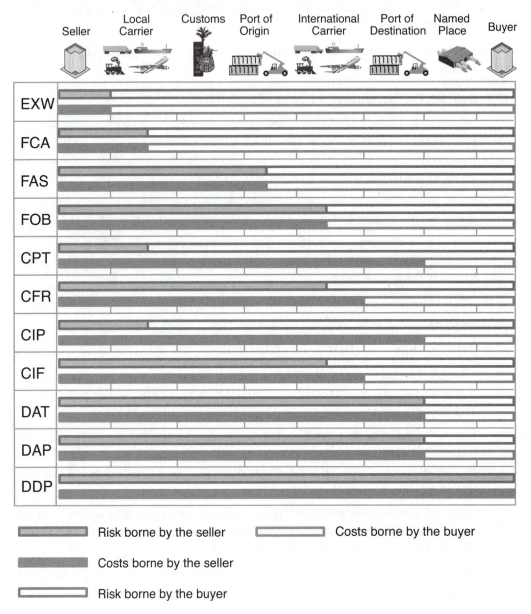

Figure 8-7a   Incoterms 2010.

Source: ICAO, Annual Report of the Council 2012.

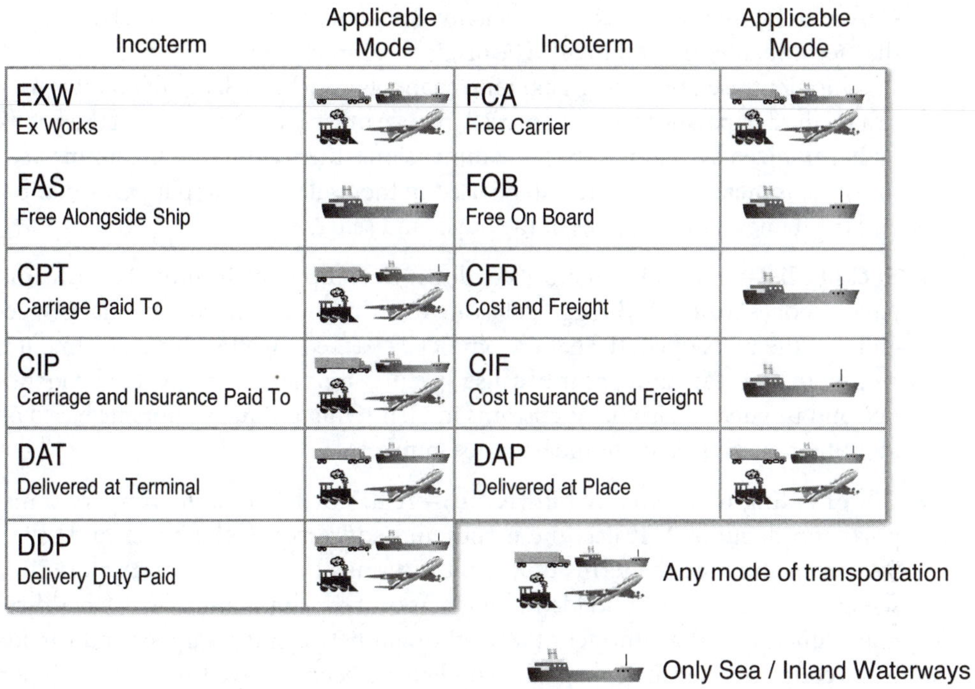

Incoterm	Applicable Mode	Incoterm	Applicable Mode
EXW Ex Works		**FCA** Free Carrier	
FAS Free Alongside Ship		**FOB** Free On Board	
CPT Carriage Paid To		**CFR** Cost and Freight	
CIP Carriage and Insurance Paid To		**CIF** Cost Insurance and Freight	
DAT Delivered at Terminal		**DAP** Delivered at Place	
DDP Delivery Duty Paid			

Any mode of transportation

Only Sea / Inland Waterways

Figure 8-7b   Incoterms 2010.

Source: ICAO, Annual Report of the Council 2012.

Key Documents Used in Enabling Transportation for Global Supply Chains

In the current environment, more countries are adopting EDI/Internet to transmit key documents among various government agencies and entities such as buyers, sellers, vendors, banks, port authorities, freight brokers, and ship owners, to name a few. However, such countries are in the minority. Most of the documentation is still done physically, and the need for accuracy and paperwork is relatively high when compared to domestic trade. This section lists some of the key documents and their functions that enable global supply chains to exit. It is by no means an exhaustive list; each country might have its own set of documents before goods are allowed to be imported or exported.

- **Pro-forma invoice**—This invoice contains the initial quote that the buyer and seller agreed to. Several Customs authorities around the world require the pro-forma invoice to be part of the mandatory documentation. The contents, prices, and terms of trade listed on the pro-forma invoice need to be reflected accurately in the actual commercial invoice and packing list.

- **Commercial invoice**—A commercial invoice (or its copy) is the actual document that accompanies the shipment. The original commercial invoice can be sent with the shipment itself or through banking channels, as when a letter of credit is used as a method of payment. A commercial invoice must state the nature of the goods, its harmonized system number, the dimensions, the weight, the Incoterms used for the shipment, the currency to be used in the trade, the shipping information, and the names and addresses of the buyer and seller.

- **Packing list**—A packing list breaks down the shipment in unit sizes (drums, pallets, boxes, bags). It disaggregates the description in the commercial invoice and lists the units of goods that are physically packed together. This document is helpful in case damages occur because specific shipments can quickly be identified and isolated. It also helps customs in identifying unique shipments based on identifiers such as batch numbers or lot numbers.

- **Bill of lading (B/L)/airway bill (AWB)**—A B/L or AWB is the primary transportation document. It lists the names and addresses of the buyer and seller, Incoterms used for the terms of trade, the terms of contract for transportation, the name of the carrier, and identifying marks of the shipment. The B/L has three main functions: a) a contract of affreightment between the shipper and carrier, b) a receipt by the carrier that the goods have been received from the shipper, and c) a certificate of the title of the goods. Normally, limited "negotiable" B/L and AWB are made of every shipment as the goods are released to the bearer of the "negotiable" B/L. A B/L is considered a "clean B/L" if the master of the vessel does not notice visible damages while loading the goods onto the ship. The B/L is considered a "soiled B/L" if defects are noted on it, which can delay payment to the seller of the goods.

- **Certificate of analysis**—Most times, the buyer wants to be reassured that the products conform to certain standards. The seller normally issues a certificate of analysis itself (or through an inspection agency) to satisfy the buyer.

- **Export/import licenses**—Certain countries prohibit or limit the import or export of certain kinds of goods. In such cases, the countries issue export/import licenses to be furnished at the time of clearance of goods at Customs. Importers and exporters need to check in advance whether the goods they want to trade are not on a "negative" or "restricted" list by any of the countries concerned and obtain any necessary export/import licenses in advance.

- **Certificate of origin**—Certain countries give least developed/underdeveloped countries preferential access to their markets at concessional import tariffs. In other cases, countries might want to trade in certain commodities with limited nations. In all such cases, the importer must furnish a certificate of origin from the exporter to prove that the goods were actually made in the country of the exporter. In most cases, the chamber of commerce in the exporter's country issues this document.

- **Certificate of inspection**—This is similar to the certificate of analysis. The importer requests that an independent agency inspect the goods either at the place of origin or at the place of destination (or both) to certify that it fits the description of the goods and adheres to the terms of the contract in weight, description, number of units, and quality of the goods. The independent inspection agency then issues a certificate of inspection with or without any defects noted on it.

- **Certificate of insurance**—When the goods are shipped on a CIF or CIP basis, the importer normally requires the exporter to furnish a certificate of insurance from an insurance agency to cover the risks of the shipment in case of unforeseen circumstances. Insurance is normally given to goods or services that have pure risk (that is, the probability of a loss only), as opposed to speculative risk (the probability of a loss or gain). Losses can be due to undesirable movement of cargo during its transportation, theft and pilferage, exposure to weather, acts of God, piracy, and *force majeure* clause. As far as shippers go, the higher the risks that are covered, the greater is the premium paid for the insurance. Types of insurance follow:

  - **Institute Marine Cargo Clauses (A)**—The highest insurance that a shipper can get is Coverage A, which is similar to "All Risks Coverage" and covers losses or damages to the shipment under consideration.

  - **Institute Marine Cargo Clauses (B)**—Coverage B is normally called the "named peril" policy because it specifically states the perils (such as fire, collision, water damage, sinking, jettison, and piracy) that are covered for a shipment.

  - **Institute Marine Cargo Clauses (C)**—Coverage C provides the minimum coverage required under the CIF/CIP Incoterms. It normally covers fire, stranding, sinking, collision, and jettisoning.

Many other types of insurance coverage exist, and insurance companies can negotiate with exporters and importers to tailor coverage to specific clients' requirements.

Summary

This chapter focused on how transportation enables the functioning of global supply chains. We covered the basic modes of transportation and their importance to the functioning of international trade. We looked at key players who participate in the transportation process. We detailed Incoterms that have standardized the terms of trade throughout the world. Finally, we examined some of the key documents that form the basis of every shipment.

Key takeaways from this chapter include:

- Understand the various modes of transportation to enable global supply chains

- Distinguish between the various kinds of sea and air services, based on type of service and size.

- Know and apply all the 11 Incoterms appropriately.

- The role of the key enablers in transporting goods in international trade.

- The use and importance of the key documents on which transportation is based, specifically the role of bill of lading (B/L).

Endnotes

1. World Trade Organization (WTO), "Understanding the WTO: Who we are," www.wto.org/english/thewto_e/whatis_e/who_we_are_e.htm.

2. United Nations Conference on Trade and Development (UNCTAD), "Review of Maritime Transport 2012," http://unctad.org/en/PublicationsLibrary/rmt2012_en.pdf.

3. National Ocean Policy Coalition (NOPC), "Oceans Impact the Economy," http://oceanpolicy.com/about-our-oceans/oceans-impact-the-economy/.

4. Admiralty and Maritime Law Guide, International Conventions, "Convention on a Code of Conduct for Liner Conferences," Geneva, 6 April 1974, www.admiraltylawguide.com/conven/liner1974.html.

9

TRANSPORTATION AND SUSTAINABILITY

Sustainability is often thought of as a negative concept forced upon businesses that increases their overall costs without any consideration to the survival of the business. Sustainability has also been wrongly associated with environmental protection. In reality, sustainability seeks to find the balance among three seemingly disparate aspects of every business: 1) securing economic value for shareholders, 2) protecting the environment in which the business operates, and 3) expressing concern for the community and other stakeholders of the business. These three aspects are rightly called the *Triple Bottom Line (TBL)*—economic, environment, and societal dimensions. Sustainability as an idea was enunciated most famously in the United Nation's Brundtland Commission report, where it was defined as the ability "to meet the needs of the present without compromising the ability of future generations to meet their own needs."[1] Agenda 21 of the United Nations, a follow-up to the Brundtland Commission, mandated that business and industry reduce their impact on the environment "through more efficient production processes, preventive strategies, cleaner production technologies and procedures throughout the product life cycle, hence minimizing or avoiding wastes."[2]

Businesses conventionally are run using a "cradle-to-grave" strategy. A firm normally offers a product and ensures that its supply chain has the capability to deliver the product to the end consumer. The firm does not monitor or have a stake in following up with the consumers on where, how, and why they dispose of the product after they stop using it. In most cases, the product ends up in a landfill, which has huge environmental costs in the long term. In today's environment, pressure from the government, consumers, and society is forcing firms to rethink the way business needs to be conducted. Firms are looking at reintroducing raw materials into the supply chain and using innovative methods to design, manufacture, and deliver products to the consumers. This is leading firms to adopt the "cradle-to-cradle" strategy. As Samuel DiPiazza, Global CEO for PricewaterhouseCoopers, puts it, "[L]eading global companies of 2020 will be those that provide goods and services and reach new customers in ways that help to address the world's major challenges—including poverty, climate change, resource depletion, globalization, and demographic shifts. This is not just about social responsibility, but about developing a core business operation that can thrive in a different global economic environment." Organizations that are resistant to change will find themselves forced to change due to legislative action such as the End-of-Life Vehicles (ELVs) Directive, the Restriction of Hazardous Substances (RoHS) Directive, and the Energy using Product (EuP) Directive.[3]

Therefore, to embrace the concept of sustainability, firms must view the world through the lens of the TBL. Transportation is one enabler of moving products from the source of raw materials to the final consumer and back from the consumer either to a landfill or to the supply chain. In addition, transportation is a derived demand and, hence, uses inputs such as services and products from other industries whose output is normally a service. The different modes of transportation can be broadly divided as personal cars, trucks, rail, air, sea/inland waterways, and pipelines. These modes sometimes are in competition with one another and sometimes complement each other. Hence, to analyze sustainability in transportation, the individual modes must be studied within the context of the supply chains in which they operate.

Transportation's Role in Sustainable Supply Chain Management

A *supply chain* is a complex web of organizations trying to match the end demand of consumers with all the supply constraints of these organizations. A focal manufacturing company could have multiple-tier suppliers and customers, each of which needs to be connected to the others to ensure a smooth flow of products. The role of transportation is to connect all the suppliers with all the customers. In addition, with the recent focus on cradle-to-cradle strategy, the role of transportation has only increased in the area of reverse logistics. Whether the end products are landfilled or reintroduced back into the same supply chain or a different supply chain, transportation enables these functions.

Two dominant views of supply chain management exist: the functional viewpoint and the process viewpoint. The functional viewpoint as popularized by the Supply-Chain Council's Supply-Chain Operations Reference (SCOR) has the elements of *plan, source, make, deliver, and return.* In Figure 9-1, transportation plays a big part in sourcing, in delivering, and in the returns of the product.

In the process view of supply chain management, transportation plays a big part in some key processes: a) managing demand by techniques such as just-in-time (JIT) and lean logistics, b) fulfilling orders in the frequency and priority that a firm desires, c) enabling the flow of materials in the manufacturing process that is normally spread out across the globe, and d) making sure that the returns process occurs efficiently and effectively. Hence, whether firms view supply chain management as a function or a process, transportation offers the physical connectivity among members of the supply chain.

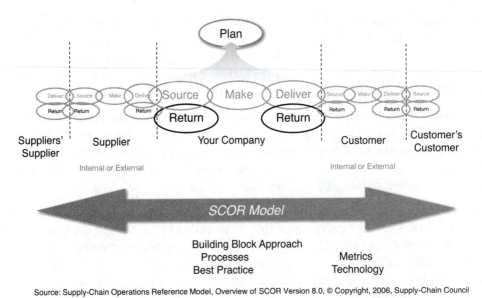

Figure 9-1 Supply chain operations reference model.

The Need for Sustainability in Transportation

Transportation accounts for approximately 19 percent of global energy use and emits about 23 percent of the energy-related carbon dioxide (CO_2). In addition, the transport sector is responsible for 60 percent of the world's oil demand, with road transport accounting for 80 percent of the oil demand of transportation.[4] Because transportation—and, hence, CO_2 emissions—are strongly correlated with population and incomes, the trend is for increased use of fossil fuels and greater emissions, which leads to an unsustainable mix. This is because the least developing and developing countries are moving toward a higher standard of living while the developed countries are in the process of maintaining, if not improving, their current levels of prosperity. Since 1971, the transportation sector has more than doubled the amount of energy it uses. The top users of energy are road (passenger), road (freight), world marine bunkers, domestic aviation, international aviation, pipeline transport, rail transport, inland and coastal navigation, and other miscellaneous modes. Within the different kinds of fuels used worldwide, North America, the Middle East, Australia, and Japan primarily use gasoline, followed by diesel, jet fuels, compressed/liquefied natural gas (CNG/LNG), electricity, biofuels, and coal. In the rest of the world (other than Russia and its neighbors), diesel dominates the fuel mix, followed by gasoline and jet fuel. Only in Russia and its neighboring countries do gasoline and CNG/LNG have an equal share and represent the majority of the fuel mix. The road (passenger) segment is often called the light duty vehicles (LDVs) by the

International Energy Agency (IEA). The IEA estimates that, except for Asia (excluding Russia and Japan), most of the LDVs are dominated by cars and sports utility vehicles (SUVs). Asia (excluding Russia and Japan) is dominated by two/three-wheelers. However, the growth of cars in Asia is now the highest compared to other parts of the world, especially in developed economies where their share is actually falling.

The average vehicle efficiencies as expressed by CO_2 equivalent per ton-kilometers (freight) indicates that shipping is the least polluter of greenhouse gases (GHG), followed by freight rail, road freight, and air. In the passenger segment as measured by CO_2 equivalent per passenger-kilometer, rail releases the least amount of GHG, followed by bus, two-wheelers, passenger LDVs, and air.

Table 9-1 shows the energy consumption by different modes of transportation in the United States. The most energy used (in petajoules) is by LDVs that use gasoline and diesel as the primary fuels. This accurately mirrors the trend happening in the world. LDVs are followed by freight by road (combination truck), passenger and freight airlines, water-/sea-borne trade, and rail. The only anomaly in the United States when compared to the global trend is that rail transport uses less energy compared to sea/inland waterways, possibly because of the prevalence of lengthier freight trains compared to the rest of the world.

Table 9-1   Energy Consumption by Mode of Transportation (Petajoules) for Select Years

	1995	2000	2005	2010	2011
Air					
Certificated carriers[a]					
Jet fuel	1,825	1,980	1,892	1,575	1,547
General aviation[b,c]					
Aviation gasoline	36	42	37	28	27
Jet fuel	80	138	217	204	212
Highway					
Gasoline, diesel, and other fuels					
Light duty vehicle, short wheel base, and motorcycle[d]	9,003	9,664	10,235	(R) 11,502	11,732
Light duty vehicle, long wheel base[d]	6,014	6,982	7,764	(R) 4,781	4,659
Single-unit 2-axle 6-tire or more truck	1,349	1,261	1,253	(R) 1,991	1,871
Combination truck	2,894	3,385	3,652	(R) 3,947	3,718
Bus	142	147	148	(R) 253	255

	1995	2000	2005	2010	2011
Transit					
Electricity	18	19	21	23	24
Motor fuel					
Diesel[e]	99	86	78	85	84
Gasoline and other nondiesel fuels[f]	8	3	4	19	20
Compressed natural gas	2	6	14	18	19
Rail, Class I (in Freight Service)					
Distillate/diesel fuel	509	541	600	511	539
Amtrak					
Electricity	1	2	2	2	2
Distillate/diesel fuel	11	14	10	9	9
Water					
Residual fuel oil	930	1,012	818	(R) 812	720
Distillate/diesel fuel oil	342	331	293	(R) 292	311
Gasoline	140	148	166	154	146
Pipeline					
Natural gas	762	699	635	(R) 733	744

Key: R = Revised.

[a] Domestic operations only.

[b] Includes fuel used in air taxi operations, but not commuter operations.

[c] The values for energy consumption by general aviation in 2010 are estimated values.

[d] Data for 2007-2010 were calculated using a new methodology developed by FHWA. Data for these years are based on new categories and are not comparable to previous years. The new category *Light duty vehicle, short wheel base* includes passenger cars, light trucks, vans and sport utility vehicles with a wheelbase (WB) equal to or less than 121 inches. The new category *Light duty vehicle, long wheel base* includes large passenger cars, vans, pickup trucks, and sport/utility vehicles with WB larger than 121 inches.

[e] *Diesel* includes Diesel and Bio-Diesel.

[f] *Gasoline and other nondiesel fuels* include Gasoline, Liquified Petroleum Gas, Liquified Natural Gas, Methane, Ethanol, Bunker Fuel, Kerosene, Grain Additive, and Other Fuel.

Source: U.S. Department of Transportation (DOT).

As far as GHG emissions are concerned in the United States, Table 9-2 illustrates the respective emissions among CO_2, methane, and nitrous oxide by different transportation modes. As expected, the amount of GHG emitted shows a strong correlation with the amount of energy consumed by each mode of transportation. LDVs, followed by trucks and buses, air, water, rail, and pipeline, are the big emitters of GHG.

Table 9-2 Greenhouse Gas Emissions by Transportation Modes, 1990 and 2011 (Million Metric Tons of Carbon Dioxide Equivalent)

	Carbon Dioxide		Methane		Nitrous Oxide	
	1990	2011	1990	2011	1990	2011
Highway total	1,190.5	1,470.4	4.2	1.3	40.4	14.5
Cars, light trucks, motorcycles	952.2	1,065.1	4.0	1.2	39.6	13.6
Medium and heavy trucks and buses	238.3	405.3	0.2	0.1	0.8	0.9
Water	44.5	47.4	0.0	0.0	0.6	0.7
Air	187.4	148.5	0.1	0.0	1.8	1.4
Rail	38.5	45.3	0.1	0.1	0.3	0.3
Pipeline	36.0	37.7	0.0	0.0	0.0	0.0
Other	0.0	0.0	0.2	0.3	0.9	1.6
Total[a]	1,497.0	1,749.3	4.6	1.7	44.0	18.5

[a] The sums of subcategories may not equal due to rounding.

Source: U.S. DOT.

Congestion has always been a problem in all modes of transportation, especially in urban areas. The Texas Transportation Institute (TTI) reported in 2005, "Traffic congestion continues to worsen in American cities of all sizes, creating a $78 billion annual drain on the U.S. economy in the form of 4.2 billion lost hours and 2.9 billion gallons of wasted fuel."[5] The Federal Highway Administration (FHWA) of the U.S. Department of Transportation (DOT) estimates that the exceeded capacity of the National Highway System (NHS) will rise from 3.35% in 2002 to 25.6% in 2035. Figure 9-2 shows the severity of the problem in all the metropolitan areas of the United States. This would lead to increased consumption of fuel and higher emissions of GHG, in addition to lost hours of work and sizable opportunity costs. More congestion also points to higher rates of accidents and potential for injury.

One of the problems leading to highway and roadway congestion is the way the Highway Trust Fund is funded. The current model pays a fixed amount per gallon of gas used into the fund. However, due to more efficient vehicles and the fact that people are driving shorter distances than they used to, there is real concern regarding the monies available to maintain the current road network. Future expansion of the highway and road system needs considerably higher funding than what the fund currently holds.

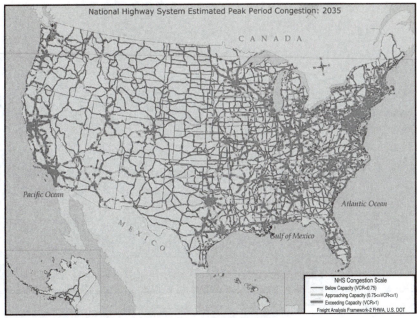

Figure 9-2   NHS highway congestion for the year 2035.

In the United States, the railroads are privatized, except in the case of Amtrak, which has the mandate of moving only passengers. As previously noted, the miles of tracks drastically reduced by 50 percent from 1960 onward, to around 140,000 miles. The rail industry saw steady consolidation after the Staggers Act in 1980, when the industry was deregulated. As of today, only seven Class I operators exist, severely limiting competition. The railroads are showing signs of investing for the future. But the Federal Highway Administration is predicting a 30 percent rise in congestion from 2002 to 2035. Figure 9-3 essentially points to a rail network that exceeds its capacity in most of the main routes connecting the West Coast to the East Coast.

As far as the congestion in airports is concerned, the U.S. Government Accountability Office (GAO) states in its report that, by 2025, 14 airports would be severely congested even if additional funding were made available to them because of location issues. Table 9-3 lists the 14 airports and their respective metropolitan areas. Even though the number of airports is small, the cascading effect of delayed flights, missed connections, and higher penalties on airlines throughout the system will be felt widely.

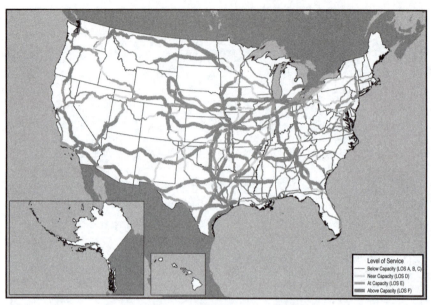

Source: Federal Highway Administration, U.S. Department of Transportation

Figure 9-3   Train volumes in 2035, compared to capacity in 2002.

Table 9-3   Airports Forecast as Being Significantly Capacity-Constrained by 2025 (Even If Planned Improvements Occur) and Their Corresponding Metropolitan Regions

Airport	Metropolitan Region
Hartsfield–Jackson Atlanta International	Atlanta
Midway International	Chicago
Las Vegas McCarran International	Las Vegas
Long Beach–Daugherty Field	Los Angeles
John Wayne–Orange County	Los Angeles
Newark Liberty International	New York
John F. Kennedy International	New York
LaGuardia	New York
Philadelphia International	Philadelphia
Phoenix Sky Harbor International	Phoenix
San Diego International	San Diego
Oakland International	San Francisco
San Francisco International	San Francisco
Fort Lauderdale/Hollywood International	South Florida

Source: Government Accountability Office (GAO), U.S. Department of Transportation.

Triple Bottom Line (TBL)

The notion of the TBL is attributed to John Elkington.[6] The TBL consists of three dimensions of business performance: profit (economic), planet (environmental), and people (societal) bottom lines. It is often represented as shown in Figure 9-4, with sustainability being the intersection of the TBL. The motivation of coming up with the three bottom lines was that improvements could be made only when they were measured accurately. Among the criticisms of TBL is that it is hard to measure. Most times, the economic bottom line can be expressed in a common denominator, such as money/currency as, say, dollars ($). But expressing environmental and societal measures for a diverse range of industries with the same units is difficult. Also, the question of allocating weights to each of the bottom lines can vary dramatically, depending on the firm and the context in which it operates.

Figure 9-4   Representation of the Triple Bottom Line.

One organization that has tried to measure TBL across different organizations is the Global Reporting Initiative (GRI), based in Amsterdam, The Netherlands. This organization has strategic associations with the United Nations Global Compact, United Nations Environmental Programme, Organization for Economic Cooperation and Development, and International Organization for Standardization. In addition, the organization has synergies with The Earth Charter Initiative, UNCTAD, and International Finance Corporation. GRI encourages firms to voluntarily report the measures based on TBL and is currently in the G3.1 version of its reporting metrics.[7] The firm then grades organizations based on the amount of disclosure on their sustainability metrics as given in their G3.1 guidelines. Approximately 5,604 organizations have started voluntarily reporting their sustainability measures to this organization (see http://database.globalreporting. org/).

The TBL was designed to be used at the firm level. Each of the bottom lines can be further broken down into different categories, but for the sake of simplicity, the following classifications are used.

Economic Bottom Line: This bottom line can be further broken down as per the various business dimensions:

- Cash flow and growth-management measures
- Balance sheet (asset utilization) measures
- Productivity and efficiency measures
- Demand management measures
- Capitalization measures

Environmental Bottom Line: This bottom line can be further broken down, based on the environmental impact areas:

- Material use measures
- Energy use measures
- Solid residue measures
- Liquid residue measures
- Gaseous residue measures

Societal Bottom Line: This bottom line can be further broken down, based on the stakeholders the firm impacts:

- Suppliers
- Financial institutions
- Customers
- Employees
- Local community
- Non-government organizations (NGOs)/media
- Government

One big debate centers on the drivers of change in TBL. The consensus so far is that if an organization or country sees benefits (economic) or if society demands a change (societal), only then will the environment change for the better. By itself, environmental concern will not force governments or organizations to reduce or eliminate GHG, emissions, or any other pollutants.

The GRI reporting guidelines mentioned earlier normally apply at the organizational level. Measuring the TBL of the entire transportation industry is difficult for these reasons:

- **Lack of common standards**—Every mode of transportation has its own set of metrics and performance standards. Often they do not complement each other. Also, the goods and passengers in certain modes of transportation cannot be easily switched to another mode of transportation.

- **Limitation of the data collected**—Multitudes of agencies are collecting different datasets that are not easily comparable. In some countries, due to war, strife, or natural disasters, data collection might be thoroughly absent. In collecting data for the entire world, a lag of a couple years often passes before good data are made available. In the meantime, the data might not accurately reflect the ground realities.

- **Different objectives**—Different governments and agencies have different objectives while running their transportation systems. Some concentrate on profitability, some on accessibility in terms of reach, and others on subsidizing the network to enable low cost access. Each of these objectives leads to different metrics being developed and emphasized.

Potential Solutions to Make Transportation More Sustainable

Every product that is manufactured has four different stages in its lifecycle: premanufacturing, manufacturing and distribution, use, and post-use. Most concepts that deal with the idea of sustainability think of it as prolonging the *use* stage of a product's lifecycle. Extending the *use* stage of a product's lifecycle only increases the *sustainment* of the product, not its *sustainability*. To increase the sustainability of the product, all four stages of the lifecycle need to be considered holistically.[8] In addition, at the premanufacturing stage, products need to be designed while keeping in mind that the raw materials to be used in the goods need to be disassembled and processed in the post-use phase of the product, to be reintroduced as raw materials in the second lifecycle of the raw material. Hence, every raw material that goes into a product needs to be looked at as having applications beyond one lifecycle of the product, into multiple lifecycles of the product. As an example, car engines are being built with aluminum even though it is more expensive than steel because aluminum can be melted and reused several times, unlike steel, which tends to break down after three or four cycles of remelting.

To be truly sustainable, the virgin raw materials introduced into the product for the first time should have the capacity of being fully or partially reintroduced into the supply chain at the end of their current lifecycle. This requires innovative design capabilities at the premanufacturing stage of the product lifecycle. It has been shown that such innovation has led to increased profitability for firms.

The end-of-life directive in Europe came into force on September 18, 2000. The directive stated that car manufacturers were responsible for the final disposal of cars after the *use* phase of their lifecycle. The directive mandated that at least 85 percent of the car be recycled. This led to a change in the composition of cars at the premanufacturing stage to enable the car manufacturers to meet the directive. Specifically, the use of plastics has increased by 50 percent to 133 kg. Thus, use of aluminum has increased dramatically, to 210 kg per car, a 120 percent increase from 2003, because that material is easily recyclable. Materials made of natural fibers are being tested to replace polymers, to increase recyclability and decrease the overall weight of the vehicle to increase fuel efficiency. Car manufacturers are increasingly using lifecycle assessment tools at the design (premanufacturing) phase to make sure that they comply with more stringent regulations.[9]

The IEA estimates that GHG emissions will be reduced in the future through these means:

- **Modal shift**—People preferring high-speed trains for short- and long-distance traveling. Because of the high price of fossil fuels, people tend to stay closer to their workplace and might choose other forms of transportation, such as walking or biking within the downtown areas of the city.

- **Efficiency**—Better ways of managing transportation networks and fleets, and newer, faster routes found over sea, air, and land to reach destinations quicker while emitting lower GHG.

- **Alternative fuels**—Electric vehicles, hybrid vehicles, and hydrogen fuel cells becoming cheaper over the years, driving up efficiency while reducing GHG.

On the policy-making front, several laws are forcing the transportation industry to be more sustainable. In the United States, the Energy Policy and Conservation Act started the Corporate Average Fuel Economy (CAFÉ) regulations to improve the fuel efficiency of cars and trucks. In 2011, President Obama signed an agreement with 13 car manufacturers to increase fuel efficiency to 54.5 miles per gallon in models starting in 2025. The EU has a nonbinding goal of cutting the average CO_2 from passenger vehicles to 95 grams per kilometer in 2020 (from 159 grams in 2007). The industry has taken the lead on reducing its carbon footprint and emissions by switching shipments to alternate modes of transport and improving tire and aerodynamic technologies in its fleet.

Transport-related CO_2 emissions are expected to rise by 57 percent worldwide from 2005 to 2030. China and India will contribute about half this rise, as their economies grow. Figure 9-5 gives the amount of CO_2 being emitted by various modes of transportation per ton of freight carried per kilometer. Air transport emits the maximum GHG per this measure, followed by road transportation and then the shipping industry. The Kyoto Protocol excludes international aviation and maritime transport from the GHG emission targets of signatory countries and lets the International Civil Aviation Organization (ICAO) and the International Maritime Organization (IMO) come up with their own guidelines. According to the IMO, the shipping industry contributed around 3.3 percent of all global emissions in 2007, with international shipping contributing 2.3 percent of all CO_2 emissions. However, no laws mandate emission cuts in the shipping industry because member states of IMO have no consensus among themselves. Currently, bigger ships that can handle 18,000 to 20,000 20-foot equivalent units (TEUs) are being built to run at slower speeds to increase fuel efficiency and reduce emissions.

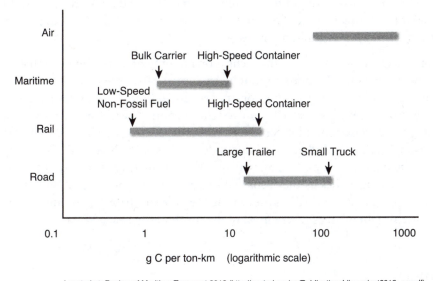

Located at: Review of Maritime Transport 2012 (http://unctad.org/en/PublicationsLibrary/rmt2012_en.pdf)

Figure 9-5 Comparison of CO_2 emissions in freight transport by mode of transport (grams carbon per ton freight carried per kilometer).

In addition, port operations are seeking to become more environmentally responsible. The Port of Gothenburg in Sweden provides a good example. The port is fitted to equip ships with shore-connected electricity generated by wind power just off the coast. This allows ships to run on cleaner power. The port is also developing liquefied natural gas (LNG) as a cleaner alternative source than fossil fuels. Finally, the port is establishing more connections with Scandinavian cities by rail, as opposed to trucks. In 2012, the port reduced CO_2 emissions by 50,000 tons.[10] These developments are representative of the actions taken by the most progressive shipping ports around the world.

In the air transport business, ICAO has set up a committee to look at reducing GHG and other pollutants, but most of the initiative comes from the local community or the industry. Many countries have stopped aircraft operations at night, to reduce noise pollution. The European Union and the United States have banned certain airlines and planes for not adhering to safety standards and preventive maintenance. The industry is looking at consolidating freight and shipping in full container loads, or bypassing air transport altogether, if receipt of the product is not time-critical.

To tackle road congestion, especially in the United States, a proposal invites the private sector under the public–private partnership (PPP) model to finance new roads. This would lead to a greater number of toll roads for a pay-as-you-go model. The critique of this model is that citizens might be unwilling to pay for roads that they had not previously paid for and might adopt a "not in my backyard" (NIMBY) attitude to prevent land from being taken away to build highways.

In the air transportation business, the focus is on building or expanding regional airports in cities where the main airports cannot be expanded. Also, the focus is on integrating different modes of transport to give consumers more choice. For example, the State of California is building a high-speed train system to compete with the airlines on the San Diego-to-San Francisco corridor. Similar rail initiatives are being planned in the states of Washington, Wisconsin, Indiana, and Illinois, as well as in the Northeast Corridor. After years of reducing capacity on the rail network, the private Class I railroad operators are expanding their capacities as they sense an opportunity in the congestion that plagues the highway system. But NIMBY is preventing the railroads from expanding aggressively in dense metropolitan areas where most of the profitability lies.

Summary

By its very nature, the transportation industry impacts the living conditions on our planet. Transportation is the biggest contributor to GHG emissions and consumes enormous sums of energy. Much of the initiative to reduce emissions and improve fuel efficiency comes from citizens who are demanding safer, cleaner, and more efficient transportation. Government legislation is among the most impactful influences on sustainability in transportation, yet some carriers are leveraging sustainability as a basis of competition by actively measuring and reporting emissions and energy usage. Railroads, in particular, are speaking of their immensely more efficient means of moving large volumes over long distances than truck transportation. In turn, rail carriers in North America are placing more emphasis on intermodal transportation and investing tens of billions of dollars in expanded capacity. Many large trucking companies are allying with the railroads to leverage these environmental benefits and to offset the driver shortage problem faced in trucking. Unfortunately, these initiatives are currently primarily limited to the most advanced markets of North America and Europe. Developing countries remain focused on simply moving goods at the lowest cost through conventional means. However, many companies have realized that employing higher orders of concern for the environment and society also favorably impacts the economic bottom line. In other words, it makes sense to be lean, green, and sustainable for a healthier business and society. Perhaps this belief can gain greater acceptance around the world in the coming years.

Key takeaways from this chapter include:

- Transportation is essential to business and personal mobility, yet its larger impacts must be considered in business and policy decisions.

- The Triple Bottom Line (TBL) concept provides a means to measure and assess impact across three dimensions of sustainability performance: 1) economic, 2) environmental, and 3) societal performance.

- Lifecycle assessment helps to evaluate the environmental impact of business decisions throughout the lifecycle of a product, including the premanufacturing, manufacturing and distribution, use, and post-use stages. The number of expected lifecycles in which materials will be used and reused is also important to consider.

- Concerns for sustainability can influence the choice of transportation mode and the selection of carriers operating within a mode.

- Industry and governments must work together to devise transportation policies that ensure sustainable outcomes.

Endnotes

1. United Nations, "Report of the World Commission on Environment and Development," General Assembly Resolution 42/187, 11 December 1987.

2. United Nations Conference on Environment and Development (UNCED), 1992.

3. T. Engen and S. DiPiazza, *A Broader Approach to Accountability,* World Business Council for Sustainable Development, 2005.

4. *World Energy Outlook 2008,* International Energy Agency, www.iea.org/media/weowebsite/2008-1994/WEO2008.pdf.

5. David L. Schrank and Timothy J. Lomax. *The 2007 Urban Mobility Report.* Texas Transportation Institute, Texas A&M University, 2007.

6. John Elkington, *Cannibals with Forks: The Triple Bottom Line of the 21st Century Business* (Oxford: New Society Publishers, 1998).

7. The GRI separates out governance structure from societal measures.

8. F. Badurdeen, D. Iyengar, T.J. Goldsby, J. Metta, S. Gupta, and I.S. Jawahir, "Extending Total Life-cycle Thinking to Sustainable Supply Chain Design," *International Journal Product Lifecycle Management* 2009;4(1-3):49–67.

9. Jason Gerrard and Milind Kandlikar, "Is European End-of-life Vehicle Legislation Living up to Expectations? Assessing the Impact of the ELV Directive on 'Green' Innovation and Vehicle Recovery," *Journal of Cleaner Production* 2007;15(1):17–27.

10. Port of Gothenburg, www.portofgothenburg.com/About-the-port/Sustainable-port/.

Section 3

Transportation in 2013 and Beyond

10

THE FUTURE OF TRANSPORTATION

The transportation needs of society evolve with time to reflect changing customer tastes, technological advances, and geopolitical/legislative climates. This chapter investigates this issue using the example of arguably one of the most successful and historic transportation channels of all time, the trans-Asian silk route (see Figure 10-1). Extending 4,000 miles across the continent of Asia and into parts of Europe, the silk route got its name from the lucrative trade in Chinese silk, which was a highly valuable commodity throughout Europe between the years 100 and 1400. Land-based transportation was achieved by moving extensive cargo on the backs of pack animals; water-based transport occurred by moving large quantities of goods on wind-powered sailboats.

Silk was obviously one of the most valuable and frequently exchanged commodities along the silk route, but several other goods were also exchanged, including pottery, paper, spices, precious stones, metals, and even slaves. Customer demand for silk and spices in the West, coupled with precious stones and metals in the East, made the route a particularly lucrative one for merchants. Over time, technological advances such as the development of caravans and caravan tracks also allowed for the trade of more varied types of goods along this route. Together with a relatively stable Mongol Empire controlling a substantial part of the route, these developments ensured that the route continued to thrive for several decades.

The Asian silk route was a great transportation network for its time. Large cities developed along its path and served as what would now be described as warehousing and cross-docking locations. Several of these cities and the associated cultures thrive to this day. However, several changes occurred in the thirteenth and fourteenth centuries that eventually brought an end to transport and trade along this route. For starters, a sociocultural shift, the growing populations along the silk route, had a deleterious effect on trade. Continuously expanding populations along the route and the associated expansion of farmland implied that merchants often had to travel through agricultural lands, arousing hostility. Skirmishes became increasingly common. In addition, the waning power of the Mongol dynasty in Europe and central Asia made it harder to sustain the silk route because security for merchants was increasingly hard to come by. Finally, in

1335, the Mongol ruler of the Middle East, Abu Said, died without an heir. His officials and officers, having easy access to weaponry and militias but being unable to agree on a successor, fought one another to stalemate and collapse. The Middle Eastern branch of the silk route closed, European traders lost contact with their Chinese counterparts, and eventually the route disintegrated.

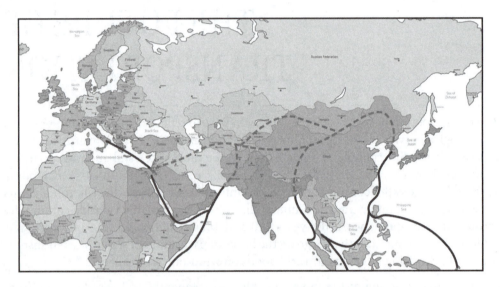

Figure 10-1   Asian silk route. Dotted lines represent land routes.
Solid lines are sea routes.

Several factors—sociocultural (the expanding population and the associated increasing private claim to land), technological (wide access to weaponry), and geopolitical/legislative (the collapse of the Mongol empire and the battles among the militias)—contributed to the paradigm shift in the viability of one of the most successful transport and trade routes of the time. Several factors such as these are relevant even today, which could affect the transportation field in the future. In this chapter, we discuss many of the emerging issues that we feel might be relevant in shaping the evolution of the transportation sector moving forward. We split our discussion into three broad trends that might have the largest impact: changes in consumer behavior (sociocultural changes), changes in technology, and changes in legislation/geopolitical factors. Some of the technologies that we have discussed are already in place and in different stages of being commercialized (some are even fully commercialized and operating profitably); others are still very early in the lifecycle and are included purely because of their *potential*.

Changes Affecting Consumer Shopping Behavior

Several new developments in the past decade either have changed or have the potential to change the way consumers shop, which eventually could affect the supply chain and, consequently, transportation needs in the supply chain. Of these, unsurprisingly perhaps, two of the potentially biggest changes are both intricately tied to one of the greatest inventions of our time: the Internet. These relate to the use of the Internet for the sale of goods (Internet retailing [IR]) and the distribution of goods (3D printing). One of these (IR) is substantially more developed than the other (3D printing) at the time of this writing, but both undoubtedly have substantial implications for the transportation industry of the future.

Internet Retailing (IR)

The year was 1994, and the first World Wide Web (WWW) browser had been released for commercial use three years prior. A young entrepreneur, recognizing the business potential of the WWW to sell goods all over the world, quit his job on Wall Street and made the nearly 3,000-mile drive to Bellevue, Washington, where he started a unique online "store." Within two months, it was selling more than $20,000 of merchandise every week. That store has today grown to be known as Amazon.com, one of the largest online retailers in the world. It enjoys revenues that exceed $61 billion.

Overall, customer adoption of the Internet for purchasing goods continues to increase at impressive rates. Online retailing has increased from just 3 percent of total U.S. retail sales in 2002 to more than 6 percent by 2008, and it is expected to account for more than 8 percent of all sales by 2014. Although many online retailers were swift to embrace the Internet for its marketing reach a decade ago, others were slow to recognize the range of challenges associated with fulfilling the grand promises of timely and efficient delivery. Nevertheless, it is now hard to find a retailer that does not have some kind of online presence. As a result, customers have many choices when they turn to the Internet to purchase goods, from online versions of their favorite department stores to purely Internet-only retailers with no physical presence.

From a retailer's standpoint, the explosion of the Internet for shopping purposes presents several unique challenges. Clearly, managing operations and order fulfillment to support online retailing presents challenges distinct from those of brick-and-mortar retailing, for which the only three rules that matter are still "location, location, location." The Internet eliminates from the equation the locational advantages that retailers in the bricks-and-mortar world have to compete for, by offering home deliveries of products bought online. However, this elimination of the location issue comes at a cost, which has implications for transportation.

Transportation Implications of Internet Retailing

Smaller Pack Sizes: Traditional retailers and, by extension, traditional transportation specialists have long been focusing on shipping goods in relatively large volumes. The traditional *manufacturer–distribution center (DC)–retail store* model of product distribution helped these business partners focus on large shipments (typically several pallet loads or even truckloads). The IR model, however, challenges this long-cultivated business model, forcing managers to solve their transportation challenges innovatively. As would be expected, as pack sizes (and, consequently, order dollar values) become smaller, transportation costs as a function of the sales price keep increasing (because of diseconomies of scale). Moreover, in the home delivery of online purchased products, most of the delivery costs are variable,[1] so there are few (if any) scale economies to be enjoyed. The net result is that there are smaller profits to be had, unless transportation costs can be managed. It has been suggested, for example, that at the time, Kozmo.com was spending close to $10 for every package delivered. This, coupled with the fact that the average order size was about $15, has been blamed for the company bleeding money and eventually folding.

Shipping to Multiple Locations: Traditional retail DCs were well equipped for one configuration: receiving products from a defined set of vendors and shipping those products to a defined (and often limited) set of customers. A major fallout of IR is that DCs are now having to deal with the challenge of shipping to several hundred thousand customers (most of whom are geographically spread out). This creates a whole new level of complexity because traditional transportation-related operational elements of DCs (palletizing, loading onto trucks, and so on) become substantially different in IR, compared to traditional retail. In the United States, several calculations have shown that only New York City has enough sales concentration and population density to support a pure Internet sales delivery business model run profitably.

Additive Manufacturing/3D Printing

We have looked at how using the Internet to *sell* goods can change the transportation field in the future. Now we look at how using the Internet for the *distribution* of goods can achieve the same. Although IR has the potential to change the way transportation occurs, 3D printing has the potential to eliminate the need for transportation completely (or at least reduce it considerably). Imagine the following scenario: You wake up in the morning and go to the sink to brush your teeth. However, just as you start your electric toothbrush, the brush head snaps off. You have no spare heads at home. In such a case, you would have to either run to the store to buy a new head and then try to get back, brush, and get dressed for work (all the while hoping that you're not late to work), or order a new head online and use an old manual brush for the day (assuming that you have one around the house). 3D printing promises to change all that. With 3D printing, you have the ability to "generate" a new brush head from scratch on your home printer. Given the right kind of printer, software, blueprints, and input materials, with 3D printing, a customer would have the ability to "manufacture" several products right at home.

3D printing (also called *additive manufacturing*) is a concept that has been around for some time. Only now, however, is the idea gaining large-scale interest. The basic principles behind 3D printing are as follows: A user designs a product, typically with a computer-aided design (CAD) program. Software then takes virtual blueprints of the CAD design and "slices" it into digital cross-sections. The easiest way to understand this is to visualize it (see Figure 10-2). Imagine slicing a cucumber into extremely thin slices. The "printer" then uses these "slices" to successively use as a guideline for printing. The printer starts from the innermost slice and builds it. It then layers the second slice on the first slice and binds it with a specialized binder. Successively, material and binders are deposited on the build bed or platform until the layering is complete and the final 3D model has been "printed." The typical home printer uses color boxes. By contrast, a 3D printer uses "matter boxes" (such as iron and carbon). This is just an extension of how the book marketplace has changed over the past decade. In the early twentieth century, buyers had to buy physical copies of books, but now they can just download the soft copy of a book. If they choose to print the book, the home printer "slices" the book page by page and prints one page (layering of the slices).

As 3D printing evolves, you can imagine online repositories where people (and manufacturers and retailers) would deposit their designs for other users to use (similar to how customers upload reviews on products or how online retailers deposit products on virtual store shelves). The Internet can thus become a source for not only the *sale*, but also the *distribution* of goods. The technology for much of what we have described already exists, but it is currently too expensive and untested to be widely adopted (see Figure 10-3 for an image of a real 3D printer). Nonetheless, there is a distinct likelihood that both of these issues will be solved in the near future and that 3D printing will become widely accessible.

Transportation Implications of 3D Printing

As expected, 3D printing promises to change the way companies do business, along several dimensions. Customers, for example, can become manufacturers and can fabricate products themselves. They will simply need the raw materials (for example, plastic pellets), which they can then use to fabricate product according to their own designs. Traditional manufacturing as we know it will see a substantial reduction in demand. Manufacturers will become more focused on developing high-quality product blueprints, which will be stored and delivered via cloud computing (especially for simple products). This will have several implications for transportation:

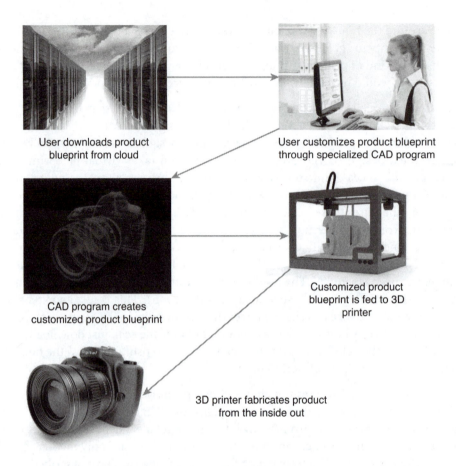

User downloads product
blueprint from cloud

User customizes product blueprint
through specialized CAD program

CAD program creates
customized product blueprint

Customized product
blueprint is fed to 3D
printer

3D printer fabricates product
from the inside out

Figure 10-2   How 3D printing works.

Reduced Need for Transportation: The ability of customers to serve as captive fabricators for themselves will reduce the need to buy manufactured goods. Of course, this will not mean that no consumers would ever buy any finished goods; for example, there will continue to be a market for physical books, even though consumers can download e-books (the equivalent of the 3D printing "blueprint") and print them (the equivalent of the 3D printing "manufacturing"). However, it is conceivable that the need for transportation will reduce at least somewhat because many nonfood items will potentially be manufactured at the end customer level.

Changing Product Mix in Transport: Currently, most downstream transport (from the manufacturer to the end customer) consists largely of finished goods. With the ability of customers to make their own finished goods, the focus of such transportation will arguably shift to transporting raw materials rather than finished goods.

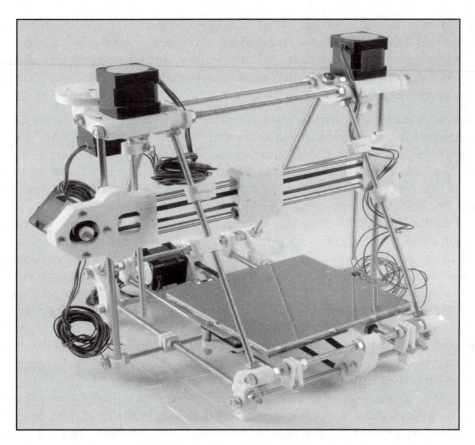

Figure 10-3   Real 3D printer.

Changes in Technology

Changes in technology are affecting not only the way customers shop, but also how transportation supports these activities. The first decade of the twenty-first century was the decade of software. Several developments in software technology occurred that directly impacted the transportation industry. These include ERP, MRP, TMS, routing and scheduling, and cloud computing. In the coming years, several other technological developments look poised to enter the mainstream, with strong implications for transportation.

Self-Driven Vehicles

The concept of the self-driven automobile has been around for several years, but it has heretofore rarely generated enough interest to be considered a viable, practical possibility.

There are several reasons for that, but we do not get into them here. The development of interest now is that the technology is becoming advanced enough that it can be considered a viable production possibility. As of this writing, three states in the United States (Florida, California, and Nevada) have all passed laws making self-driven vehicles street-legal within their boundaries.

A few technologies are currently trying to compete for supremacy in the self-driven car marketplace. However, all of them have some basic elements in common: They all involve some combination of location monitoring (such as GPS—see Chapter 6, "Transportation Technologies"), radar, and laser/lane visualizing. The GPS system helps the vehicle find out its current location, its destination, and the most efficient route to get there. Preloaded maps with posted speed limits ensure that the vehicle stays under the speed limit at all times. The laser/lane visualizing equipment captures a moving view of the road ahead. The digitized image is parsed for straight or dashed lines, the lane markings. The vehicle is supposed to stay within the lines at most times, so if it deviates and approaches or reaches the lane marking, the vehicle nudges itself away from the marker, just as a human driver would do. Finally, radar equipment ensures that the vehicle does not come unduly close to other bodies (such as vehicles and pedestrians). The combination of these three technologies working together and communicating with each other by way of an onboard computer allows the vehicle to drive itself (see Figure 10-4).

Transportation Implications of Self-Driven Vehicles

The large-scale introduction of self-driven vehicles in the mass market is merely a few years away. Several commercial manufacturers have been either developing or testing self-driven vehicles in a big way. Manufacturers such as General Motors, Daimler, and Nissan have publicly stated that, by the year 2020, they expect to have several commercially available automobile models that can drive themselves. Google has set a more ambitious target of 2017 for commercial introduction of this technology in the marketplace. Although the previously mentioned applications are largely for self-driven cars, a similar level of excitement is emerging around self-driven trucks. Caterpillar Inc., for example, already introduced a fleet of self-driven 240-ton trucks at an iron ore mine in Australia. There are several implications of this for the transportation industry, some of which we discuss next:

- **Reduced labor costs**—The United States alone has an estimated 5.7 million licensed professional drivers, and they drive the country's vast fleets of delivery vans, trucks, and tractor-trailers. According to the *Wall Street Journal*, a full-time driver with benefits costs the transportation company around $65,000 to $100,000. Eliminating this need for labor eliminates a substantial cost from the transport network, which will result in better margins and lower landed costs of goods. Of course, not all the drivers will be replaced with robots in the near future, but some fraction seemingly will. With the estimated driver shortage of 15 percent or more across the United States, the introduction of self-driven trucks promises to provide much relief to transportation managers and directors.

- **Changing route-planning algorithms**—Currently, many truck operators drive along fixed routes, with trucks returning to their homes every few days (in some cases, even every day). Thus, most trucks really operate as a back-and-forth shuttle type of service between a set of locations. However, with the advent of self-driven vehicles, the need to return a vehicle to base will be greatly reduced, especially because drivers will not need to go back home periodically. Trucks can be sent to places where they are most needed, thus creating efficiencies in transportation network optimization.

- **Hours of service (HOS) rules**—Currently, no regulations cover any HOS rules surrounding self-driven vehicles, but this will not necessarily be the case after the technology is introduced in the mainstream. Given that the trucks can be driven through automation, though, there seems to be no direct reason to impose any HOS restrictions. This means that the amount of idle time can potentially be reduced, thereby lowering transportation costs and increasing the shelf life of landed products.

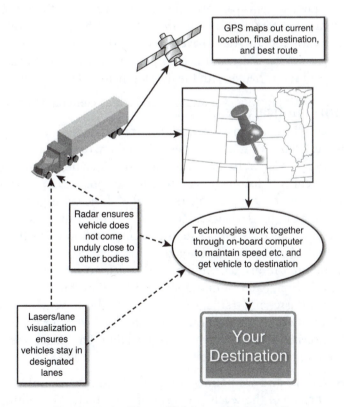

Figure 10-4   How self-driven vehicles work.

Intelligent Transport System (ITS)

Imagine the following scenario: You are driving to the ballpark to watch your favorite team play in the biggest rivalry game of the year. You are running late. When you get there, you realize that the parking lot is full (no surprises here!). Now you begin looking for parking; it takes you 30 minutes to find a spot, and another 15 to walk to the field. You are not happy, and understandably so. Your friends tell you, "You should have gotten there earlier." Now imagine if your car could guide you to the nearest open spot without your having to look around. You would save precious time, you wouldn't miss the game, and you could accomplish some more work at office before you left. That is the promise of ITS. The idea behind ITS is to connect every vehicle in a network of transportation users in such a way that each vehicle instantly tracks and shares information (the key is sharing). When all (or at least a substantial number of) users report information to the centralized ITS, everyone can quickly determine where the accidents and tie-ups are and what routes to take to avoid them. For the freight transport, this translates into quicker drive times by way of more efficient traffic patterns and less time the freight "sits on the road."

Many modern cars and trucks today come equipped with mobile positioning and navigation systems. These systems can provide turn-by-turn directions and information on, say, the restaurant, coffee shop, or rest area. Many of these systems also incorporate real-time traffic information to warn drivers of congestion and accidents. The problem, however, is that these systems communicate with the driver, but not with other vehicles or the road. This will soon change with the development of ITS.

Transportation Implications of ITS

Several governments have recognized the transportation implications of having an integrated ITS in place. China, for example, spent $2.8 billion on ITS in 2009. The U.S. government's ITS Strategic Plan 2010–2014 also envisions "a national, multi-modal surface transportation system that features a connected transportation environment among vehicles, the infrastructure and passengers' portable devices." It goes on to forecast a "connected environment [that] will leverage technology to maximize safety, mobility and environmental performance." To this end, the Transportation Department is investing in a number of new transportation technologies, including $11.5 million for vehicle-to-vehicle safety communications; $9.3 million for vehicle-to-infrastructure communications, such as traffic signal and timing updates; $2 million for real-time data capture and management for assessing traffic, transit, and freight movement patterns; and $8 million for dynamic mobility applications to find optimal ways for people and goods to be transferred between different modes of transportation. The biggest implication of ITS will be traffic management and waiting times:

- **Reduced transit and wait time**—Proponents of lean manufacturing consider waiting to be one of the seven key wastes to eliminate. Yet waiting is prevalent.

The average American spends four days every work year just waiting in traffic. According to the 2012 Urban Mobility Report published by Texas A&M University, the "cost" to trucks by waiting in congestion (in terms of fuel, wasted time, and so on) was $27 billion in 2012 (not including the value of goods being transported, spoilage charges, and so on). At present, companies such as Google collect real-time traffic data through crowdsourcing—if a user has Google Maps installed on a mobile phone with GPS capabilities enabled, that person can transmit his or her location to Google in real time, to determine what road the car is on and at what speed he or she is traveling. When Google combines this information with similar information gathered through other phones on the road, across thousands of phones moving around a city at any given time, it gets a reasonably good idea of live traffic conditions. However, because vehicles are not directly reporting data into a centralized grid, the information currently being gathered is rather spotty. (For example, if four people are traveling in the same car, from a traffic standpoint, it is only one car—from Google's standpoint, however, it is four people, and Google has no way to tell the difference.) Thus, the implementation of ITS has the potential to save freight transit time substantially—to the extent of $30.2 billion, according to some estimates.

Geopolitical, Legislative, and Societal Changes

The past two decades have seen several changes in the geopolitical climate of the world. Rising affluence levels in many parts of the world, the fall of the Iron Curtain, and the breaking down of barriers have impacted transportation in unique ways. These and other changes will continue to change many facets of the freight business. In this section, we discuss a sampling of the changes the freight business is experiencing.

Rising and Erratic Fuel Prices

The past decade has seen fuel prices rise all across the world. In the United States alone, the retail price of gasoline has risen from an average of about $1.50 per gallon in 2003 to about $3.50 per gallon in 2013, an increase of more than 130 percent. During the summer of 2008, the nationwide average surpassed $4.60 per gallon for diesel fuel. This steep increase in prices stems from several factors. One of the largest is that, with increased globalization, income levels in the poorer countries of the world are rising rapidly. These rising income levels have created a stronger world demand for fuel (through increased usage of personal automobiles, longer commutes, and travel for leisure). With the supply tightly controlled by a few producers, the overall effect has been to push prices up. As affluence levels in the developing world continue to increase, the demand for fuel will continue to increase. This indicates that unless alternative energy sources are employed, fuel prices will continue to head upward for the foreseeable future. This upward pressure on fuel prices has important implications for the transportation industry.

Transportation Implications of Rising Fuel Prices

Many freight businesses and governments have recognized the inevitability of the upward trend in fuel prices. As a result, several initiatives are expected to impact the transportation industry in the future. Some of these are driven through political action; others are industry driven:

- **Fuel-efficient fleets**—In many countries (including the United States), the move to enhance fuel efficiency of the transportation fleet is two-pronged, with efforts from both government and industry. The federal government, for example, has proposed to impose fuel efficiency standards on trucks based on weight and intended use. For example, over-the-road tractor-trailers would be required to achieve a 20 percent reduction in fuel consumption and CO_2 emissions by 2018. Heavy-duty pickups and vans would be subject to different gasoline and diesel fuel standards, with reductions ranging from 10 to 15 percent. Other work trucks would have to reduce fuel consumption and greenhouse gas (GHG) emissions by 10 percent by 2018. Several older tractor-trailers thus would have to be replaced in the near future in favor of more fuel-efficient ones, implying upcoming capital expenses for many companies.

- **Alternative-fuel vehicles**—Alternate fuels have been making their way into automobiles for some time now, but most of the development has taken place for cars rather than commercial vehicles. Several challenges arise when adapting the same technology for commercial vehicles, especially trucks. A recent alternate-fuel development in motor cars technology is to make the vehicles all electric. However, this is a challenge for tractor-trailers because of the amount of freight they need to haul, and even more for the massive distances they need to cover. Battery technology has just not developed to the point that such massive amounts of power can be stored without compromising massive amounts of space. To tow two trailers, the first trailer would have to be all batteries, which is a highly unviable solution. The same problem exists in outfitting trucks to run on compressed natural gas (CNG) because CNG requires about six times as much storage space as diesel, even when squeezed down to 3,000 pounds per square inch. This is a serious waste of storage space on vehicles whose primary purpose is moving large volumes of goods. However, an alternative fuel is now emerging: liquefied natural gas (LNG). LNG is created by chilling natural gas to 260°F below zero and squeezing it down 600 times in volume, and it needs only about 1.7 times the amount of storage space as diesel. Upon startup, such trucks still use a few squirts of diesel to get going, but overall, diesel use typically gets cut by about 95 percent. Currently, the bottleneck for wider adoption technology is the lack of fueling stations, which runs about $1.5 million. However, it is only a matter of time before more widespread adoption of this development takes place.

Global Warming and Greenhouse Gases

The term *global warming* refers to the rise in the average temperature of the earth's atmosphere and oceans since the late nineteenth century and its projected continuation. Since the early twentieth century, the earth's mean surface temperature has increased by about 0.8°C (1.4°F), with about two-thirds of the increase occurring since 1980. Substantial consensus within the scientific community is that one of the major causes of this phenomenon is the increased concentration of GHG produced by human activities such as burning fossil fuels and deforestation.

Transportation Implications of Global Warming

The deleterious effects of global warming have been well publicized (rising water levels, submerged low-lying areas, reduction of the polar ice caps, and so on), and the transportation implications of the same are clear (for example, the move to alternative-fuel vehicles). However, the unintended positive effects haven't received much attention:

- **Opening the Northern Sea Route**—Global warming has transformed the Arctic in recent years, and its summer ice cover has dropped by more than 40 percent over the last few decades. Some news reports indicate that the Arctic could be completely free of ice by the summer of 2030. This raises the prospect that it might soon be possible to sail along the Arctic's sea routes with ease. This is an idea that is proving irresistible to shipping companies, mining companies, and oil and gas exploration companies. The Arctic route can cut 12 to 15 days (indicating a savings of 25 to 30 percent in terms of travel time) from traditional routes between Asia (especially China) and Europe.[2] In a world where time is money, this savings in transit time is a goldmine for ocean shipping companies. The rise in shipping interest along this route is staggering: In 2010, only four ships sailed this route; in 2011, this number increased to 46. In 2013, at least 370 vessels will likely sail this route, a growth rate of 9,250 percent in just 3 years!

The Need for Infrastructure to Support Growing Populations

In 2011, the world's population passed a milestone by exceeding 7 billion people. In 1928, the world's population stood at merely 2 billion. The United Nations estimates that the population will rise to somewhere between 7.4 billion and 9.2 billion by 2030.

(Other sources put this number at 8 billion.) The sharp increase in population, coupled with rising living standards around the world, will stress transportation and logistics systems already under duress. A related trend to population growth is increasing urban migration, particularly in rapidly developing countries. More people are moving away from the agricultural economy in pursuit of opportunities in the cities. This migration worsens congestion and further stresses urban infrastructure.[3]

Transportation Implications of Population Growth and Migration

Some nations are investing heavily in their infrastructure to accommodate increased volumes. China, for instance, reportedly has invested the equivalent of 9 percent of the nation's gross domestic product (GDP) in transportation infrastructure, with approximately $1 trillion directed toward investments in cargo infrastructure. Most of this investment is focused on the nation's export capacity at coastal shipping locations. Transportation remains quite challenged inland to the west of these port locations. In fact, it is believed to cost more to ship freight from western cities in China to Shanghai than to ship across the ocean from Shanghai to Los Angeles.

Other nations are finding it difficult to finance dramatic investments in transportation. The United States, for instance, is facing a crumbling infrastructure among many of its modes. Those that rely on public funding (roadways and water transportation), in particular, are facing battles to maintain and extend the infrastructure to sustain the growth in freight and passenger traffic. Many locks and dams used to support barge transportation on the Mississippi and Ohio River corridors were built to last 50 years; many are still in operation after more than 80 years of heavy use. From time to time, disasters such as the Interstate 35 West bridge collapse in Minneapolis, Minnesota, in 2007 serve as sober reminders of the need to maintain existing infrastructure. As of 2012, more than 10 percent of bridges were rated as "structurally deficient" in the United States, with another 13 percent regarded as "functionally obsolete" to support modern traffic.[4] Clearly, solutions are required to support safe and efficient transportation of people and goods.

Without sufficient public funds, many local, state, and federal governments are turning to private funding sources. *Public–private partnerships (PPPs or P3s)* are joint ventures between the public sector and one or more private commercial interests. The partnerships assume a variety of investment formats, but they were designed to support the new development or enhancement of existing infrastructure that previously was regarded as a public asset. Understandably, the commercial entities have a profit motive, which strikes some as disconcerting for a "public asset" to be used to generate commercial interests. Partnerships are employed in toll roads, bridges, and parking structures, among other forms of infrastructure. PPPs are on the rise throughout much of the world, particularly in locations with urgent investment needs for infrastructure development.

Increasing Demands for Security and Safety in Transportation

As noted early in the text, transportation is an essential activity for any economy. The more developed the economy, the higher the level of trade and, hence, the more dependent the nation finds itself on reliable transportation. To actively participate in the economy, it is essential for people and goods to employ transportation. People and freight share many common rights-of-way, such as roads. The accidental injury and death statistics tell the unfortunate, tragic story inherent to people and heavy equipment interfacing on roadways.

In addition, our dependence on transportation can make transportation modes a target for groups who seek to exert coercion through violent means and disrupt an economy. Terrorists find transportation assets and movement activity to be rich targets for invoking fear, interfering in people's lives, and interrupting the flow of goods. The disparate nature of transportation assets and the overwhelming volumes of traffic make it virtually impossible to guarantee safe transit throughout the world.

Transportation Implications of Security and Safety

With the deregulation of economic matters in transportation throughout much of the world, regulatory focus is shifting to matters of safety and social concerns. In the United States, for instance, HOS regulations for drivers of commercial motor vehicles remained unchanged from 1962 to 2004. The rules changed again in 2013 as regulators learned more about the role of fatigue in contributing to accidents. Ongoing research will likely result in more changes in the future, including the requirement for carriers to use electronic on-board recorders (EOBRs) to monitor driver time, replacing the paper logbooks that drivers have used for many decades. Similarly, individual drivers, and the carriers that employ them, are held to a higher standard for safety performance through the Compliance, Safety, and Accountability (CSA) program initiated in 2010. Emphasis will continue along these lines, with expectations and accountability for safety increasing.

As for terrorism, the primary focus to date remains the safe transport of passengers. More concern will be directed to cargo in the future. An immediate concern is the prevention of cargo transportation as a means to willfully deliver harmful devices and chemicals. Documentation and monitoring of suspicious shippers serve as the primary means of detecting potentially dangerous cargo—only a small share of cargo containers arriving at seaports is x-rayed or inspected upon arrival. Surveillance technologies that allow for faster, efficient screening of cargo remain under development and will see deployment at the largest ports in the near future.

Another security concern that will grow in importance is the assurance of safe transit of essential goods, such as food products. It is imperative that the security of a nation's food supply remain intact throughout the supply chain, including the time in transit. Greater attention will be placed on assurances of proper containment to protect consumers from in-transit food tampering. On a related basis, agricultural shippers will find it necessary to maintain the genetic identity of the grains they sell throughout their distribution. Genetically modified organisms (GMOs), or genetically engineered crops, that help to improve pest resistance and crop yield are feared in many settings, and are even banned in the European Union (EU). It becomes essential, then, to distinguish GMO from non-GMO grains, to ensure that the two are not comingled at any point between the farm and the food processor. To date, the practice of distinguishing facilities (such as grain silos and elevators) and transportation vehicles (such as trucks and railcars) for GMO and non-GMO has not been common practice. Such requirements can influence how facility and transportation capacities are dedicated and used in the future.

Summary

After reading this chapter and this book, you know that the transportation needs of society evolve with time. Transportation (especially goods transportation) is a *derived demand,* in the sense that it is created as an outcome of other economic necessities. As consumer tastes, societal preferences, and international relations evolve, consumption patterns change—and transportation changes to reflect these developments. Making predictions on the future of the nature of the transportation business is fraught with risk. For example, a mere 25 years back, few would have predicted that the U.S.'s largest import partner in 2013 would be China, a country whose government the United States did not even formally recognize until 1979. In this chapter, we have presented some trends that lately seem to be emerging strongly: Internet-based commerce models (for example, online shopping and 3D printing), automation (such as driverless vehicles), environmental factors (as with alternative-fuel vehicles and the Northern Sea Route), growing demands for transportation, and safety and security throughout the supply chain. These issues will shape the direction the transportation business takes in the days to come. Let us all watch, see, and reflect in a decade on whether the predictions hold true.

Key takeaways from this chapter include:

- Transportation has played a significant role in the formation and development of societies, and it will continue to influence future developments.

- Technologies such as the Internet influence the ways that commerce is conducted. As e-commerce continues to grow, companies must adapt their transportation systems to economically accommodate home delivery.

- New technologies are among the most significant developments within the transportation field that are changing the way businesses think about and implement transportation management.

- Geopolitical, legislative, and societal changes will continue to exert considerable influence on transportation practices. The firms that adapt to these environmental circumstances effectively will achieve significant advantages.

Endnotes

1. See T. M. Laseter and E. Rabinovich, *Internet Retail Operations: Integrating Theory and Practice for Managers* (Boca Raton, FL: CRC Press, 2011).

2. Robin McKie, "China's Voyage of Discovery to Cross the Less Frozen North," *The Guardian*. August 17, 2013. http://www.theguardian.com/world/2013/aug/18/china-northeastern-sea-route-trial-voyage

3. Chad W. Autry, Thomas J. Goldsby, and John E. Bell, *Global Macrotrends and Their Impact on Supply Chain Management: Strategies for Gaining Competitive Advantage* (New York: FT Press, 2013).

4. *Ibid.*

For Further Reading

Ashley, S. (2013), "Robot Truck Platoons Roll Forward," www.bbc.com/future/story/20130409-robot-truck-platoons-roll-forward. Accessed 24 September 2013.

Blair, S. (2005), "East Meets West under the Mongols," *The Silk Road Foundation Newsletter,* www.silk-road.com/newsletter/vol3num2/6_blair.php. Accessed 24 September 2013.

Durmaine, B. (2012), "The Driverless Revolution Rolls On," *CNN Money,* http://tech.fortune.cnn.com/2012/11/12/self-driving-cars/. Accessed 24 September 2013.

Goldsby, T. J., S. S. Rao, and S. E. Griffis (2011), "The Opportunities and Challenges of Online Retail Logistics," *Logistics Quarterly* (Winter): 46–47.

Guizzo, E. (2011), "How Google's Self Driving Car Works," *IEEE Spectrum* (October): 11–12.

Laseter, T., and Rabinovich, E. *Internet Retail Operations: Integrating Theory and Practice for Managers,* 1st ed. (Boca Raton, FL: CRC Press, 2011).

Rao, S. S. (2012), "Internet Retailers Get Revved Up—The Logistics of Electronic Commerce from a B2C Perspective," *Industrial Management* (October): 14.

Index

Numbers

C

cabotage, 167-168

CAD (computer-aided design), 211

CAFE (Corporate Average Fuel Economy) regulations, 200

canals, 32-33

Capesize ships, 165

car bombs, 11-12

carbon footprint of transportation industry, 11

cargo air transportation, 173

cargo aircraft, 36-37

cargo types
- bulk cargo, 164
- containerized cargo, 164
- ocean transportation, 163-164

Carriage and Insurance Paid To (CIP), 182

Carriage Paid To (CPT), 182

carrier cost metrics, 50-53
- ABC (activity-based costing), 52-53
- CTS (cost to serve), 50-52
- OR (operating ratio), 50

carrier freight bill (CFB), 106

carrier numbers in different modes of transportation, 175

carrier pricing, 54-56

carrier selection, 90-93

carrier-shipper negotiations, 96

centrifugal networks, 83

centripetal networks, 83

certificate of analysis, 186

certificate of insurance, 186-187

certificate of origin (CO), 107, 186

certification of inspection, 186

CFB (carrier freight bill), 106

CFR: Cost and Freight (CFR destination port), 181

charterers, 162-163

CHAs (custom house agents), 179

CIF: Cost, Insurance, and Freight (CIF destination port), 181-182

CIP: Carriage and Insurance Paid to (CIP place of destination), 182

Class I railroads, 30

class rates, 56-57

CNG (compressed natural gas), 218

CO (certificate of origin), 107, 186

CO_2 emissions, 200-201

Code 128 bar codes, 130

cold chain packaging technologies, 139

cold chain temperature-monitoring technologies, 139-140

collaborative, planning, forecasting, and replenishment (CPFR), 150-151

collaborative transportation management (CTM), 151-154

commercial invoices, 107, 185

commodity classes, 57-58

Compliance, Safety, and Accountability (CSA) program, 221

compressed natural gas (CNG), 218

computer-aided design (CAD), 211

congestion, 194-196

consolidation
- examples, 86-90
- inbound/outbound consolidation, 86
- temporal consolidation, 85
- vehicle consolidation, 85

consular invoice, 107

consumer shopping behavior, changes affecting
- additive manufacturing/3D printing, 210-212
- IR (Internet retailing), 209-210

container ships, 31-32, 166

containerized cargo, 164

contract rates, 96-97

contractual provisions, 97-99

control and monitoring systems
- location monitoring systems, 136-137
- overview, 136
- temperature control and monitoring systems, 138-140

Corporate Average Fuel Economy (CAFÉ) regulations, 200

Cost, Insurance, and Freight (CIF), 181-182

Cost and Freight (CFR), 181

cost to serve (CTS), 50-52

costs
- of 3PLs (third-party logistics providers), 69-70
- ABC (activity-based costing), 52-53
- accessorial fees, 60-61
- accounting costs versus economic costs, 47
- carrier pricing and costs for shippers, 54-56
- CTS (cost to serve), 50-52
- demurrage, 60
- detention fees, 59-60

P

packing list, 105, 185

Panamax ships, 165

passenger air transportation, 170-171

passive shippers, 139

payments (TMS), 123

PBL (performance-based logistics) programs, 101

PCPs (pneumatic capsule pipelines), 44

performance assessment, 99-101

performance-based logistics (PBL) programs, 101

pipeline transportation, 37-38. *See also* intermodal transportation

    PCPs (pneumatic capsule pipelines), 44

    U.S. transportation infrastructure, 8-9

place utility, 3

pneumatic capsule pipelines (PCPs), 44

POD (proof of delivery) document, 106

point-to-point networks, 82

pooled distribution, 149

population growth and migration, 219-220

Post New Panamax ships, 165

PPP (public-private partnership) model, 202

PricewaterhouseCoopers, 189

pricing

    accessorial fees, 60-61

    carrier pricing and costs for shippers, 54-56

    demurrage, 60

    detention fees, 59-60

    diversion or reconsignment fees, 59

    expressions of transportation rates, 56-59

        class rates, 56-57

        FAK (freight-all-kinds) rates, 57-59

        flat rates, 56

    linehaul price, 59

    surcharges, 60

private road fleets

    advantages, 65-66

    largest private fleets in U.S., 63-64

private transportation

    definition of, 63

    nonroad private fleets, 66-67

    private road fleets

        advantages, 65-66

        largest private fleets in U.S., 63-64

Proctor & Gamble

    4PLs (fourth-party logistics providers), 70

    bullwhip effect, 113

    information sharing with Walmart, 114

pro-forma invoices, 185

proof of delivery (POD) document, 106

providers

    EDI providers, 119

    R&S providers, 128

    TMS providers, 125

public-private partnership (PPP) model, 202

Q-R

Q-max ships, 166

R&S (routing and scheduling) systems

    benefits and applications, 126-127

    implementation, 128

    overview, 125-126

    providers, 128

radio frequency identification (RFID), 133-135

rail transportation, 25-30. *See also* intermodal transportation

    advantages, 25

    efficiency, 26

    equipment, 27-28

    international transportation, 26-27

    rail market, 30

    speed, 25-26

    Staggers Rail Act, 25

    U.S. infrastructure, 25

    U.S. transportation infrastructure, 8

Rails to Trails Conservancy, 25

rate shopping, 120-121

receiver-shipper negotiations, 94-96

reconsignment fees, 59

regional railroads, 30

relay networks, 91-93

reports (TMS), 123

request for proposal (RFP), 97

requests for quotes (RFQs), 97

Restriction of Hazardous Substances (RoHS) Directive, 190

revenues, U.S. domestic modal split, 15

reverse flows (logistics), 5

RFID (radio frequency identification), 133-135

RFP (request for proposal), 97